Hydr...

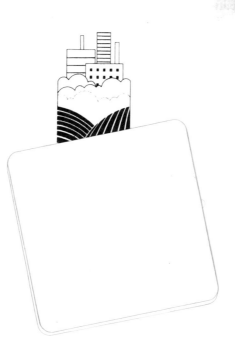

BASIL BLACKWELL

Contents

© Phillip Jones 1983

First published 1983
Reprinted 1984, 1986, 1987, 1988, 1990

Published by
Basil Blackwell Ltd
108 Cowley Road
Oxford
OX4 1JF

ISBN 0 631 12951 0 (paperback)
0 631 12941 3 (school edition)

Printed and bound in Great Britain
by Dotesios (Printers) Ltd., Trowbridge, Wiltshire

1 The hydrological cycle

Introduction

*Everything originated in the water
Everything is sustained by water.*

Goethe's sentiments quoted here from *Faust* reflect the importance of water to man and help justify the study of hydrology. Such beliefs have been expressed before and since Goethe. Vitruvius, a Roman architect and town planner, gave instructions for planning towns which included the necessity 'to describe methods of finding water . . . inasmuch as it is of infinite importance for the purposes of life, for pleasure and for our daily needs'. Much later in the United States the Mississippi Valley Committee stated that 'people cannot reach the highest standard of well-being unless there is the wisest use of land and water. Flood control, drought control, navigation, power, water supply, sanitation and erosion are integral parts of the picture'. More recently the United Nations launched a hydrological decade to stress the growing pressures on water on the planet.

The human body itself is over 75 per cent water; it can exist for periods of weeks without food yet, deprived of water for a fraction of that time, will lose control of body temperature, become comatose and possibly die. World population is increasing as is individual consumption of water. In 1929 each citizen of the United Kingdom used approximately 126 litres of water a day. By 1975 the figure had risen to 310 litres, over a third of which was used for washing, bathing and lavatory flushing, illustrating the link between growing water consumption and increasing standards of living. The 1976 drought in Britain (see figs 1.1 and 1.2) increased by 300 per cent the price of animal foodstuffs in some parts of the country; it proved more profitable to give away cattle than face the food bill to keep them. In the same year the drought threatened and brought about short time working for industry. To make one tonne of aluminium takes 1 363 000 litres of water; a family saloon motor car takes 454 000 litres; a tonne of steel 100 000 litres, and nine pints are needed to make one pint of beer!

Our demands on water resources are large and varied; they are also conflicting. Different demands for water in the United Kingdom have to be met by the

Fig 1.1   A queue for water at a street standpipe during the 1976 drought in England and Wales

Fig 1.2   Tottiford reservoir in 1976

various regional water authorities. In England and Wales the water authorities dispose of 63 414 million litres of sewage and industrial waste each day, and are also responsible for prevention of pollution, flood protection (fig. 1.3), maintenance of inland fisheries and other forms of water-based recreation. They must ensure that the supply to domestic consumers is plentiful and potable (drinkable); and they must achieve all this with tight financial control. The water authorities realise, perhaps better than anyone, the importance of the hydrologist in providing water for society's infrastructure, at local, national and international level.

The concept of hydrology as outlined in this Introduction is an anthropocentric (human-oriented) one. It focuses on meeting society's demands for water for a variety of uses.

The hydrologist becomes involved in practical investigations of a detailed and complex nature both to monitor and control the effect of his work. He has to understand how precipitation moves into, through and over the earth's surface, and the consequences of these movements. This book looks at the processes whereby water moves, the responses to these processes of rivers and their drainage basins, and the impact of man on each of these.

EXERCISE

1.1 Study figure 1.4 and answer the following questions:
(i) In what ways has man interfered with the natural flow of water in the drainage basin?
(ii) List some uses of water in an industrial economy.
(iii) Figure 1.4 shows water used for agricultural, domestic, recreational and industrial purposes. Is it possible to put these uses in an order of priority?

The hydrological cycle

The movement of water on the planet can be illustrated by the use of a model – the hydrological cycle (fig 1.5). Hydrologists study both the *atmospheric* and *terrestrial* parts of the cycle, the former because the temporal and spatial variation in such things as temperature and precipitation exert a considerable influence on the behaviour of water once it reaches the land. The terrestrial part of the cycle, namely infiltration, moisture in the ground, river channels and drainage basin patterns is that part most often affected by the activities of man, and as such is the main focus of hydrological study. This book analyses the workings of the hydrological cycle and applies the understanding gained to observe ways in which water is studied, used and controlled by man. The complexity of the hydrological cycle is being increasingly realised, not least as the costs of interfering with it become apparent and are subjected to critical political and economic analysis.

It is impossible to be precise about the total amount of water in the hydrological cycle on earth. Figures which can be quoted are meaningless to anyone with a perception of everyday volumes of water. The US Geological Survey estimates the total planet water volume to be about 1 384 000km³. Much more interesting and relevant is the distribution of this water throughout the planet, for it is this which determines man's ability to develop the earth's water potential. The pie charts in figure 1.6 indicate the accepted distribution of water within the hydrological

Fig 1.3   Urban flooding: but it's an ill wind . . .

Fig 1.4   Man's use of water

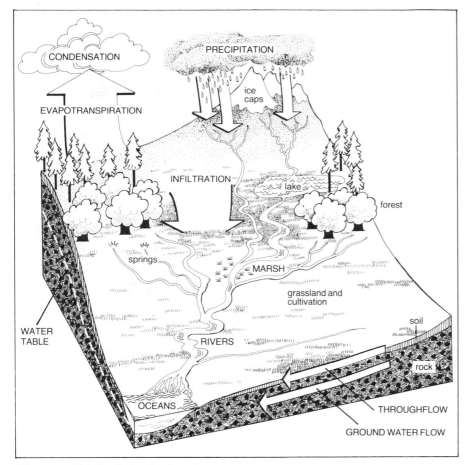

Fig 1.5 *The hydrological cycle*

EXERCISES

1.2 How would you explain the patterns of extreme drought and extreme wetness shown in figure 1.7?

1.3 Figure 1.7 is based on mean annual rainfall totals. How useful are these figures to the hydrologist?

1.4 Compare figure 1.7 with a world map of population. Which parts of the world appear to suffer the worst gap between supply and demand?

1.5 The land area of Great Britain is 229 914km². The average rainfall of Great Britain is 904mm. The population of Great Britain is 54 387 000. (All these figures exclude Northern Ireland.)
(i) What is the total volume of rainfall that falls over Great Britain? (Be careful with your units.)
(ii) How many cubic metres of water are there for each person in Britain?
(iii) Why is your answer to (ii) meaningless when related to the demand for water from individuals in Britain?

cycle, but it is important to realise that the totals indicated within any part of the cycle are not constant in time or space.

Thus, before the advent of seawater desalination, 97 per cent of the planet's water was salty and unusable in the oceans. The 3 per cent of theoretically usable water is in fact 75 per cent unusable, locked up in glaciers and ice caps, leaving only 25 per cent of fresh water (0.8 per cent of the total water) available for human use. Even this is not all suitable for all purposes, some of it being far too mineraliferous due to contact with rocks of the earth's crust; it is not always easily accessible, some

being too deep; nor is it evenly distributed over the earth's surface (fig 1.7). The hot dry tropics and the interiors of some continents far removed from the sea are very deficient in water. High latitudes may receive equally low rainfall, but are saved from aridity by low evapotranspiration rates due to low temperatures. The majority of rain falls on equatorial or mid-temperate latitudes. The latter zones house the bulk of the world's population, and hence the maximum demand for water, though this is changing with the economic development of the densely-populated areas of the tropics (fig 1.8).

At the moment, most of the water used by man is taken from surface water, i.e. rivers and lakes. Major sources of this water remain unused e.g. the Amazon discharges 175 000m³ per second at its mouth; the Zaire 40 000m³ per second, and both are largely untapped at present. The Colorado on the other hand, whose mean annual discharge is 6370m³ per second, barely reaches the sea because so much water is abstracted from it along its course for use in the dry states of Colorado, Nevada and California.

In view of these figures, the fact that the ice caps hold enough water to feed all the world's rivers at their present rates of discharge for nine hundred years makes their storage capacity incomprehensibly large. Developing these large riverine and ice cap storages is not viable at the moment. Practical and political problems make many world rivers uncontrollable as integrated units in the foreseeable future.

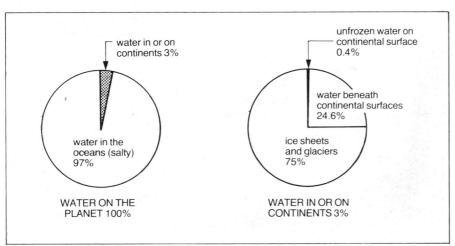

Fig 1.6 *Distribution of water within the hydrological cycle*

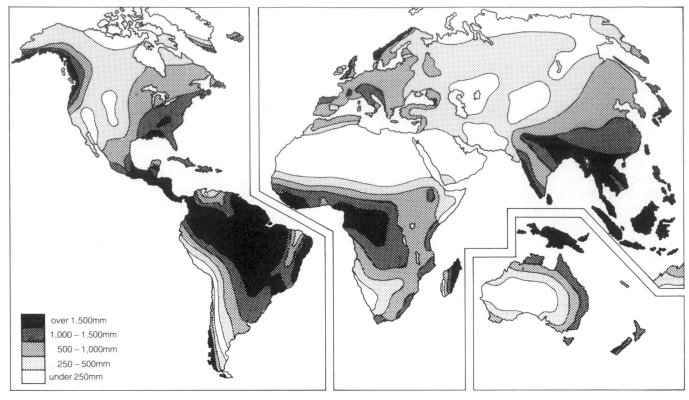

Fig 1.7   *Distribution of annual world rainfall*

over 1,500mm
1,000 – 1,500mm
500 – 1,000mm
250 – 500mm
under 250mm

American schemes to tow icebergs southwards to supply Los Angeles are equally impracticable, while the thoughts of melting the ice caps by man-induced climatic change would be cataclysmic for anyone anywhere in the world living within 30m of present sea level.

We therefore cannot expect to increase our water supply dramatically in the immediate future and as a result will be forced into more efficient use of the water resources which are currently tapped. We will be forced to discover hitherto untapped water resources within technological reach, and forced to seek more economical power resources for desalination of oceanic water.

Hydrology and the systems approach

The 1960s and 70s saw the development of a new approach to the analysis of geographical problems – the systems approach.

A system is an attempt to break down situations into their component parts, called elements, and to study the relationships, sometimes called linkages, between them. Within the context of hydrology, the drainage basin is the ideal unit in which to study water and its effects as a system.

The drainage basin, a subdivision of the earth's land surface area, is a system which receives inputs of matter via precipitation, and energy inputs via solar energy and the gradient of the land; it has an output from it in the form of water and sediment entering the sea, and also loses moisture to the atmosphere. Hydrologist R.C. Ward expressed this idea thus: 'water is at work within the drainage basin from the moment of input of a raindrop on the soil surface to its final exit from the basin in the main trunk stream'. It has already been suggested in this volume

that water is influenced by a number of events in its movement through the drainage basin and, in turn, can influence the form of the drainage basin itself. Figure 1.9 shows how the movement of water within the drainage basin can be expressed in the form of a simple system diagram. This diagram should be looked at in conjunction with Figure 1.5 and the elements of the system related to the landscape.

The value of the study of systems is increased by accurate measurements (quantification) of all manner of

Fig 1.8   *Mose-la-Tunya: the smoke that thunders – but are the Victoria Falls a wasted water resource?*

6

drainage basin characteristics. As an example of the value of such measurements the Stanford Watershed Model stands out with its success in forecasting runoff patterns from drainage basins. The model computerises precisely recorded data representing drainage basin storage units, precipitation and evapotranspiration, and then stimulates a hydrograph whose form, thanks to the computer, can be adjusted by varying the values of the inputs and transfers between storage units. Figure 1.10 shows how successful it can be when applied to a small Devon catchment area.

Other benefits of precise records and measurements can be seen in the section on floods and flood prediction (pp. 49–54).

Finally, the drainage basin can be viewed as a unit in matters perhaps beyond the scope of this chapter, namely in the realm of human geography. The vast majority of mankind lives in the lower portions of river valleys and in all manner of ways people are intimately linked with the water within their drainage basin. Frederick Le Play even suggested a drainage basin land-use model akin to the more famous one by Von Thunen (fig 1.11). This shows, in an idealised model, how water within the drainage basin is needed for an array of uses, and it is clear that management is vital to avoid the consequences of diminished flow or misuse. A further factor worth mentioning in this context is that rivers, in historical terms, have generally been seen as boundaries or barriers to movement rather than as integrating factors, and this has tended to work against the efficient management of their resources. This will be examined further in chapter 9.

EXERCISE

1.6 Choose an example of a flood plain from your atlas. What use has man made of the flood plain, and what has the role of the river been in his activities? Do you get the same answers for the developed and less-developed world?

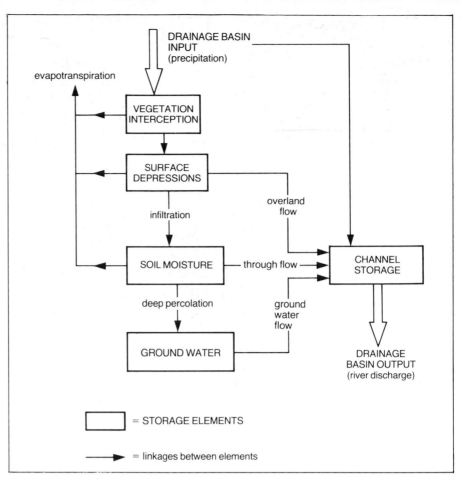

Fig 1.9   The drainage basin expressed as a simple system

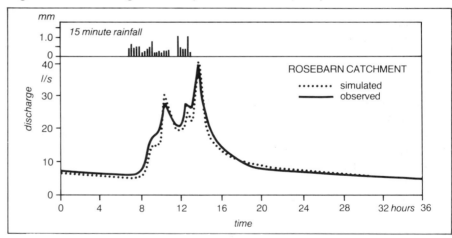

Fig 1.10   The Stanford IV Watershed model applied to a small Devon catchment area

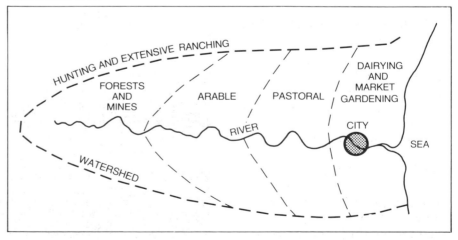

Fig 1.11   Le Play's valley section: an example of a drainage basin as a unit in human geography

2 Factors affecting infiltration

Most hydrologists become interested in the hydrological cycle when precipitation reaches the earth's surface. Precipitation may be *intercepted* by vegetation, may *run off* on the ground surface, or *infiltrate* the ground. The factors involved in determining which of these things happens to precipitation, or what proportion of a given rainstorm goes to each possible course, are involved and inter-related and fundamental to hydrology.

The significance of infiltration lies in the fact that water infiltrating the earth's surface has important geomorphological effects, particularly in producing the soluble load of rivers, while water moving over the surface of the ground affects many aspects of landform development, as well as having profound effects on man's activities (e.g. by causing soil erosion).

Nature of precipitation

This is the aspect of atmospheric hydrology of greatest interest to the hydrologist. For a given storm he will be interested in the total amount of precipitation that falls, the duration of the storm, the area it covers, and whether drops are large and heavy or small and light. A storm of small drops, low total precipitation and short duration will be largely intercepted by surface vegetation and may not even reach the soil. Conversely, a heavy storm of large drops of high intensity, prolonged over a considerable period, will react very differently on the same area. The vegetation barrier will be penetrated by so much water in such a short space of time and the soil surface may become so compacted by raindrop impact that it becomes vitually *impermeable*. Under such conditions infiltration will be low and the majority of the storm will run off the surface leading to soil erosion and flooding (see fig 2.1). A third situation is possible where rain may fall steadily, and vegetation may break up the raindrops into smaller sizes that can more readily enter the soil. Water may then move through the soil but little runs off on the surface to cause damage; the majority penetrates the soil and underlying rock where it is stored for future use by plants or man,

Fig 2.1   Torrential rain resulting in surface runoff

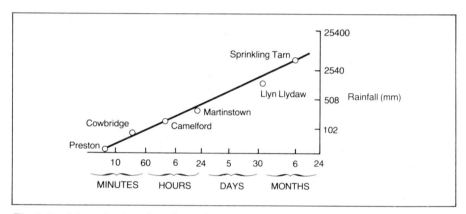

Fig 2.2   Magnitude–duration relationships for UK extreme rainfalls

or eventually flows into rivers as base flow. Finally, a consideration of rainfall intensity and its duration and recurrence interval is important (see fig 2.2). In general the more intense the precipitation the shorter it lasts, the wetter the day the less frequently it occurs, and the shorter the recurrence interval of storms the less water from each successive storm will infiltrate. In the United Kingdom high intensity rainfall is more common on the high western hills than on the east coast (see fig 2.3).

Reference to precipitation so far has assumed it to be rainfall; other types of precipitation, e.g. snow and hail, all have different effects on infiltration. For example, the rate at which snow melts will have a great bearing on how much of the melt water infiltrates. Rapid rates of snow melt, particularly if underlain by frozen ground, can cause large-scale runoff and floods.

The chemical composition of pre-cipitation is also significant because some rain-dissolved minerals can react with clays causing them to swell and bringing about compaction of the soil by reducing spaces between particles, particularly in some tropical soils.

Vegetation interception

The amount of precipitation intercepted by vegetation will depend on the nature of the precipitation and also on the type of vegetation present. The species of plant, their age, their density, the season of the year, will all be important. Given storms of equal intensity, a mature layered equatorial rain forest will intercept far more incoming rainfall than a new crop of maize planted one metre apart in rows, which will encourage greater rates of infiltration. Constant figures for vegetation interception are consequently impossible to

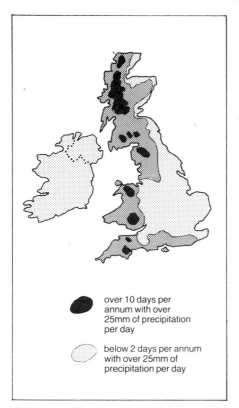

over 10 days per annum with over 25mm of precipitation per day

below 2 days per annum with over 25mm of precipitation per day

Fig 2.3 Frequency of high intensity rainfall in the British Isles

Fig 2.4 Gully erosion following forestry drainage

derive but M.J. Kirkby estimated vegetation interception for simplified computational purposes, to be approximately equal to the first 1.0mm of precipitation and 20 per cent of the subsequent storm rainfall. Work done on the subject in the United States has indicated more precisely the role of vegetation, though it *alone* does not influence infiltration.

Table 2.1

Ground cover	Infiltration rate mm/hour
Old permanent pasture	57
Heavily grazed permanent pasture	13
Strip cropped	10
Bare crusted soil	6

Pine forests, for example, are said to intercept up to 94 per cent of low-intensity precipitation but only 15 per cent of high-intensity, the average for temperate pine forests being 30 per cent.

It will already be clear that removal of the natural vegetation cover by man, followed by cultivation, often using damaging agricultural practices, can greatly reduce rates of infiltration and lead to accelerated erosion (see fig 2.4). The reverse can also be true in that the 'landscaping' of highland reservoirs in

Fig 2.5 Chequer-board forestry in an attempt to control water movement

Britain with areas of pine forest can considerably reduce the movement of water into the reservoirs; in some cases forestry practices have been modified to attempt to control water movements (see fig 2.5).

An indirect effect on vegetation in determining infiltration is through its role in affecting soil structures. Vegetation contributes enormously to maintaining a well-balanced soil structure and such 'natural' soils have a higher infiltration capacity than soils which man has cultivated for a long time. Sometimes organic material will provide an 'organic route' which water will follow when moving through the soil.

These organic routes are called *biopores*.

Depression storage

Typical British landscapes contain a myriad of surface depressions that store water on the ground thereby giving it a greater chance to infiltrate and reducing the risk of erosive surface runoff. Man's farming practices such as harrowing and rolling tend to reduce the number of surface depressions, an exception being contour ploughing where every furrow holds water kept back by the adjacent ridge (see fig 2.6).

Evapotranspiration

Varying amounts of water on or near the earth's surface return to the atmosphere by *evapotranspiration*. This is a composite term for the two main processes by which water returns to the atmosphere, namely evaporation and transpiration.

Evaporation takes place from exposed water and soil surfaces when there is a vapour pressure gradient between the atmosphere and the evaporating surface. When air is saturated, i.e. relative humidity equals 100 per cent, no evaporation takes place. Energy is needed to change water from its liquid state in seas, lakes, rivers or soil into a vapour form in the atmosphere (see fig 2.7). Nature provides this energy by solar heating and as air temperature rises so does its potential for holding water vapour, and hence greater evaporation occurs. The heat of the earth's donor surface is important, not just because it determines the temperature of the air above it but also because at higher temperatures water molecules are more active and likely to break free of the forces holding them to the earth's surface. Relatively less evaporation takes place from large water bodies than small ones because more incoming solar radiation is used in heating the water to depth and not just the surface layers conducive to evaporation. Highly saline waters have lower evaporation rates than fresh water – on average 30 per cent less in the oceans. Winds affect evaporation because they generate waves on water surfaces which increase the surface area from which evaporation can take place, and they also bring in fresh supplies of unsaturated air to maintain a vapour pressure gradient by removing air at or near saturation level. Lastly the texture and depth of soil determines the readiness with which it parts with moisture, clays being particularly reluctant to do so.

Measurement of evaporation can be done with an evaporation pan not less than 119cm in diameter and 24.8cm deep (see fig 2.8). The accuracy of results can be affected by the positioning of the pan, the addition of more water by rainfall into it, animal and bird consumption from it and so on. Eliminating these errors can lead to the creation of an artificial environment in which evaporation rates can be different from those occurring naturally nearby. Water cannot evaporate from soil if it is not there to be evaporated at the potential rate indicated by prevailing temperatures, relative humidities and wind speeds of the air above it. There is therefore often a discrepancy between potential and actual evaporation rates which is at its greatest during prolonged hot dry periods.

Rates of evaporation are strongly affected by the presence of vegetation. Plants afford shade which reduces evaporation, but plants themselves are a major element in the transfer of water to the atmosphere because of the way they *transpire*.

The majority of plants obtain their nutrients in solution via their roots and to maintain this essential flow of foodstuffs moisture has to leave the plant. It does so either directly from the tissue of plant cells or, in the majority of cases, via the stomata of leaves. This root to leaf movement of moisture is called the *transpiration stream* and its strength will depend on the ability of the atmosphere to induce water loss from the plant, and also on the nature of the soil which will determine its ability to supply water at the root level. It is obvious that the total transpiration loss per unit area

Fig 2.6   Contour ploughing: an attempt to reduce surface runoff and conserve soils

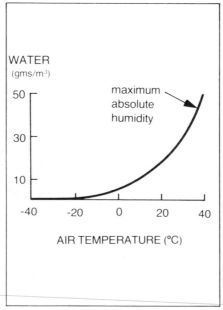

Fig 2.7   Relationship between air temperature and absolute humidity

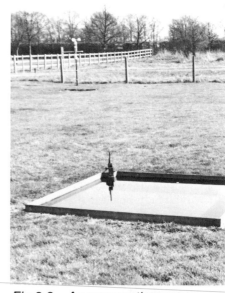

Fig 2.8   An evaporation pan

over the world will depend on the season of the year and the nature of the plant cover.

The measurement of transpiration is not easy. It has been done by botanists demanding precise measurements from single plant leaves but hydrologists want data for areas of land, perhaps extensive areas of the same crop or plant assemblage. Because of the close similarity between volumes of water lost by evaporation and evapotranspiration, the latter is usually measured by hydrologists. This can be done either practically using a lysimeter (see fig 2.9) or empirically based on the links between evapotranspiration and measured characteristics of the atmosphere such as net radiation at the earth's surface or mean daily windspeed. One of the most accurate of the latter methods is called Penman's formula, and it bears out closely lysimeter readings taken over the same period, with the advantage that it can be computerised for ease of application. On the basis of the Penman formula it is possible to produce maps such as the one shown in figure 2.10.

By compiling graphs of precipitation, potential and actual evapotranspiration it is possible to arrive at a soil moisture budget (fig 2.11) which, among other things shows the amount of water available for plant growth at different times of the year, and indicates the need

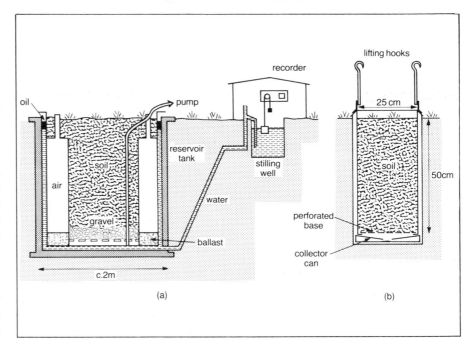

Fig 2.9   (a) hydraulic and (b) weighing lysimeters

for supplemental irrigation. This is a major area of application for hydrological study, and some specific cases are studied in chapter 9.

Nature of the soil

Soils have two major characteristics which influence the amount of water which can enter or flow through them:
1 *permeability* is the ease with which water flows through soil and rock;
2 *porosity* which is the percentage of soil volume occupied by voids (pores). These two factors are affected by a range of physical characteristics. *Soil texture* is very important; a coarse-grained sandy or gravelly soil with a low clay

content will allow water to move through it rapidly, but in soils where the clay content is much higher the pore spaces between the particles are smaller making water movement through the soil much slower. It is doubtful if even well-drained clay soils have an infiltration rate in excess of 20–30 per cent that of sandy soils. Soil texture will also determine the amount of moisture retained in the soil, which in turn affects how much more water the soil is capable of admitting. Clays are much more water-retentive than sands, and so have a lower potential for water intake. Musgrave and Holton illustrated this with some work in Illinois in 1964 (Table 2.2).

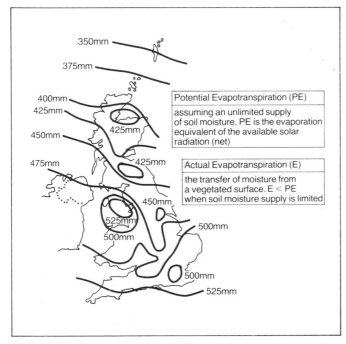

Fig 2.10   The mean annual potential evapotranspiration surface derived by using the Penman formula

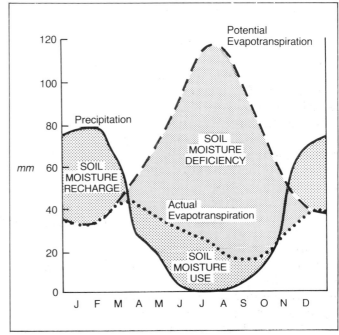

Fig 2.11   Soil moisture budget for Los Angeles

Table 2.2   The effect of initial soil moisture content on infiltration rates

	Infiltration rate mm/hr	
Initial moisture content %	*Good grass cover*	*Poor weed cover*
0–14	18	6
14–24	7	4
24+	4	3

Other soil characteristics of importance would be soil depth, which can affect the total capacity of the soil, and which can produce different infiltration rates at different depths as the layers of soil become compacted by the weight of layers above them; and secondly the presence within soils of large voids or *macropores* (fig 2.12). Recently a lot of attention has been paid to the effect of macropores on infiltration rates and it seems that they develop more readily in some soils than others and act as channels along which water can move rapidly and in volume, either horizontally or vertically, within the soil. They can originate from rabbit burrows, plant root development or dessication cracks (particularly likely in clays). They are also found where the subsoil geology is highly-jointed as in some basalt rocks, though the best example is probably the fractured lithology of limestones (fig 2.13). These macropores are more likely to persist for long periods if the soil or rock containing them is cohesive like clay which doesn't collapse to fill them in. Their presence in most soils means that infiltration can take place rapidly along them, and more slowly through the matrix of soil particles forming the rest of the soil, what Weyman calls the 'two-phase infiltration process'. The slower type of infiltration may be affected by water which is stored on the surface; such water is unlikely to be there if underlain by a macropore. Urban areas have interesting parallels with macropores; urban surfaces are mostly impermeable and to remove surface water extensive drainage systems exist whose drains could be the equivalent of macropores in natural environments.

Slope factors

Slope angle will affect the availability of surface storage depressions and types of vegetation, and in general terms the steeper the slope the lower the rate of infiltration. Slope length is also important because on a permeable soil the longer the slope the more chance surface water has to penetrate the soil as it moves downslope. In the case of an impermeable soil, the reverse is true, and runoff will *increase* with length of slope. These features of slope will obviously have no effect if the ground is already saturated.

A more useful term under which to consider the influence of slope is the texture of dissection of an area; the more dissected an area, the more badland-like that it is, the smaller are the chances of infiltration and the greater the surface runoff.

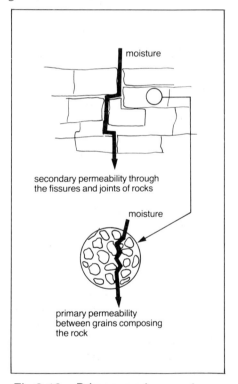

Fig 2.13   Primary and secondary permeability in rocks. The same types of movement can be identified in soils

Conclusion

The factors which determine rates of infiltration into the soil are fundamental to the rest of hydrology because they are part of the initial transfer and absorption of atmospheric moisture into the earth's surface. To attempt to quantify the rates of infiltration is diffi-

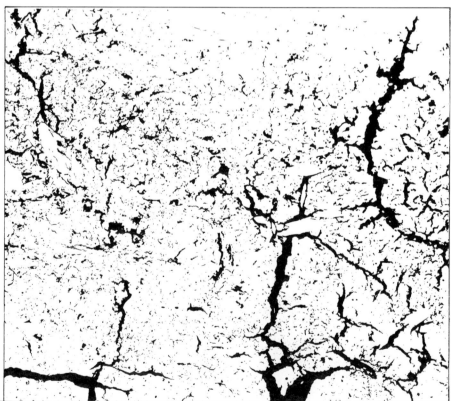

Fig 2.12   Pore spaces in soil (shown in black). The elongate planar pores, e.g. top right, are important for soil drainage, producing a passage for the flow of water through the soil. (Horizontal section, frame length 20 cm.)

cult, but current preference for the throughflow models stresses that infiltration rates can decline through time.

More frequently hydrologists try to correlate infiltration with isolated factors affecting infiltration. Fig 2.14 shows some of these relationships in a generalised form.

EXERCISES

2.1 Plot the locations given below on an outline map of Great Britain and try to account for the variation in the data given.

Table 2.3

Place	Mean annual runoff as a percentage of mean annual rainfall
Fort William	85
Berwick-on-Tweed	55
Dumfries	65
Sheffield	35
Ebbw Vale	75
Colchester	15
Isle of Wight	24

2.2 (i) What happens when rain falls around your school or home?
(ii) Under what conditions does water lie on the surface of gardens and parks?
(iii) Which sort of rainfall fills the gutters and drains of your home town; a summer thunderstorm or steady winter rain? Attempt a brief explanation of your answers.

2.3 Study figure 2.14. Describe and explain the relationships shown in each of the graphs by writing a paragraph on each of them. What practical use might a hydrologist make of them?

2.4 Study the figures given in Table 2.4 for a soil moisture budget, and then attempt the questions following. You may also find it helpful to refer to figure 2.11.

Table 2.4

	J	F	M	A	M	J	J	A	S	O	N	D	Total (mm)
Precipitation	18	18	20	21	40	58	43	34	30	15	19	20	336
Actual evapotranspiration	0	0	0	27	56	76	63	45	34	18	2*	0	321
Potential evapotranspiration	0	0	0	40	82	115	138	118	65	35	0	0	593

(All figures in mm)
*An apparent anomaly which arises because of the way P.E. is calculated.

(i) Using an arithmetic y axis, draw a graph of the data shown.
(ii) During which months is soil moisture recharged?
(iii) During which months is there a soil moisture deficiency?
(iv) Which three months have the greatest soil moisture defficiency? Put them in order and state the defficiency in mm.
(v) During which months are (a) the greatest, and (b) the least amount of soil moisture used, in addition to the month's rainfall, to make up actual evapotranspiration?
(vi) Which of the three following areas does the data come from: West African Savannalands, Brittany (France), the Canadian Prairies? Give your reasons.

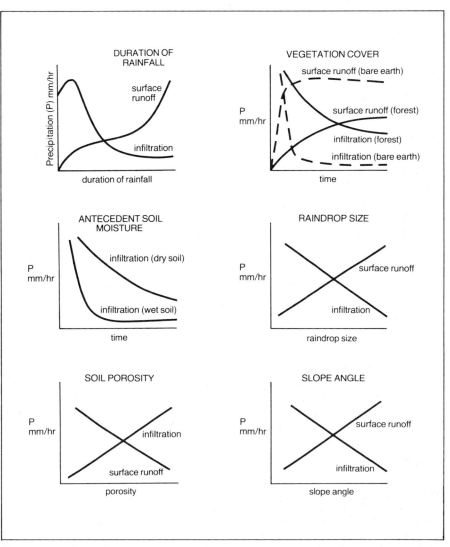

Fig 2.14 Simplified illustrations of some factors affecting infiltration and surface runoff

3 Water in the ground

Soil moisture

Water moves into the ground in amounts and by methods determined by the infiltration factors mentioned in chapter 2. Its behaviour once it enters the soil is affected by the affinity, or lack of it, of soil particles for water, in the same way as condensation nuclei attract moisture during the formation of raindrops. Within the soil, the movement of moisture is largely influenced by the force of gravity, but it is also affected by capillary and molecular forces. Three zones of soil moisture can be defined.

The effect of gravity is to produce downward movement of moisture through the soil until, at depths which vary from soil to soil, the inter-particle spaces become filled with water and the soil is therefore *saturated*. The upper limit of this saturated zone marks the limit of the *water table*. Acting against gravity is capillary suction where water moves upwards through the soil in the narrow spaces between soil particles to produce a zone of partial saturation, sometimes called the zone of *capillary saturation* (fig 3.1). Within this zone and above it water will exist as a thin film on the surface of individual soil particles, held there by molecular bonding so strong that water held in this manner is often inactive hydrologically. In all air spaces within the soil water may exist in *vapour* form.

It has been shown above that the ease with which water can move through the soil will depend on the size of the spaces between particles which will determine the soil's permeability. Thus, coarse sands are more permeable than fine clays but the latter encourage capillary action. In practice, however, in really fine clays the rate of capillary movement can be so slow that it never reaches an equilibrium state with rainfall and evaporation.

Water zones within the soil are rarely horizontal and water will move through soil along lines of greatest potential gradient. The speed at which it moves will depend on the slope of this gradient but is, of course, also subject to the macropore movements mentioned above. Rapid rates of infiltration will not necessarily produce rapid rises in the height of the water table. If sub-surface geology is highly permeable,

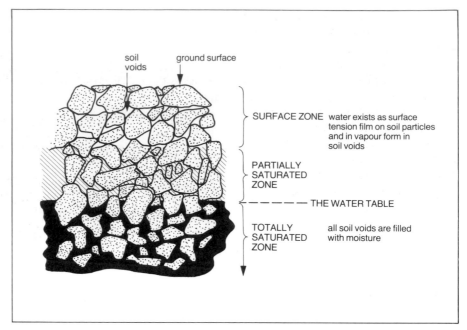

Fig 3.1   Soil moisture states

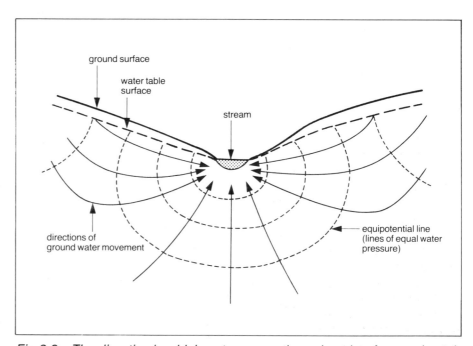

Fig 3.2   The direction in which water moves through rocks of approximately uniform permeability into stream channels

e.g. well jointed limestone or basalt, then the water table height will change slowly. Even as its height increases there will be a corresponding rise in the 'head' of soil moisture, which will lead to a more rapid movement of water through the soil, which will in turn tend to lower the water table again. Directions of water movement beneath the water table are not obvious but an idealised form is shown in figure 3.2.

An interesting application of soil moisture knowledge has taken place at Eilat in Israel (fig 3.3). Eilat in the arid Negev desert has been transformed into a garden city by using highly saline water (up to 14 000 parts per million) for direct irrigation of sand dunes. The feared accumulations of saline material at the surface did not occur because the salts were flushed downwards through the deep dune sand by regular applications of water. Water moving rapidly down through the soil left sufficient water vapour in the soil pores to sustain a variety of commercial and ornamental plants which make the settlement more attractive to those living there. It is unlikely that the scheme would have proved successful had the soil contained a greater percentage of clay than the 2 per cent it did contain, because the salts being applied would have reacted with the clays, causing them to expand and reduce permeability, and at the same time ensured a permanent harmful salt presence near the root layer of plants. The possibilities which this scheme suggests for increasing the food production of other arid zones of the world are obvious, and similar schemes have been attempted in desert areas in India.

Ground water

It has already been suggested that the sub-soil geology may receive water from the water-infiltrated soil layers above them. Virtually all geological types are permeable and porous enough to retain water in some degree. The figures in Table 3.1 indicate just how great the range can be.

As in sandy soils the permeability of solid geology will be affected by the size, shape, uniformity and method of bonding of its component particles.

Similarly, water will move through solid geology under the same set of laws that govern its movements through soil. There may exist here, too, zones of total saturation (the water table) and partial saturation. The profile of the water table will more or less follow the pattern of the surface relief, more so in low permeability rocks than in those of higher permeability. This sort of water is 'ground water' and will be a major and fairly constant source of water which moves into streams (base flow), and a steeply sloping water table will, given equal permeability, produce a faster movement of water into stream channels than a gently sloping one (fig 3.2). The level of the water table will vary from season to season and in chalk areas can produce seasonal streams such as the famous 'winter bournes' of Dorset. A reverse movement of water is also possible during prolonged dry periods, water moving *from* the stream channels into the solid geology (fig 3.4).

Ground water supplies much of the water used in the world at the present day and new sources of it are eagerly sought. Famous examples are the artesian wells of Australia, the aquifers

Fig 3.3 The garden city of Eilat (Israel). Note the arid hillsides beyond the city

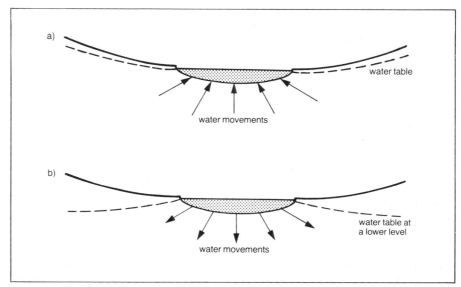

Fig 3.4 Ground water movement (a) a large hydrostatic pressure from a high water table causes water to move into stream channels (b) a low hydrostatic pressure from a reduced water table may cause water to move from the stream channel into adjacent ground, particularly if the latter is fairly porous

Table 3.1 Geological variations in porosity and permeability

Geological type	Porosity (% void space)	Approximate permeability (litres/day/m²)
Unconsolidated alluvial gravel	25–30	460 800 .0–46 080 000.0
Limestone	0.1–10	0.0046–460.8
Sandstone	5.0–25	0.046–4608.0
Basalt	0.001–50	0.0046–46.08
Unweathered granite	0.0001– 1	0.0000046–0.00046

which support the citrus production of southern California, or the water currently drawn from beneath London. It is, however, frequently overlooked that the recharge of such water sources is slow and the water table can be progressively lowered by excessive extraction rates. Consequences of such actions are dire: salt water is invading the Californian wells; London is having to dig ever deeper wells which not only proves costly but can also lead to well capture (fig 3.5). It is no wonder that attempts are being made to recharge London's aquifers artificially by pumping water *into* them.

15

Surface runoff versus throughflow

The way in which rainwater moves into stream channels, apart from direct precipitation into them, has been a major subject for hydrological debate. The undoubted classical theory was that of Horton who advocated *overland flow*, but in recent years his theory has found less favour than theories of *throughflow* or *interflow*, particularly so following laboratory experiments and field measurements which, world wide, show that infiltration rates are much higher than the maximum precipitation intensities commonly encountered.

Horton's model suggests that the majority of water moving into stream channels does so over the surface of the ground following a period of such intense rainfall that infiltration does not absorb it all. That this does occur there is no doubt and it has been observed happening particularly in arid areas. Here rainfall, often torrential, falling on an unvegetated surface compacts the surface by washing fine particles into the spaces between coarse particles at the surface thus impeding drainage and permitting surface runoff to occur very soon after the beginning of the storm even over low angle slopes. This suggests that vegetation is crucial to the Horton theory, as indeed it is: a dense vegetation cover will intercept precipitation and assist infiltration to such an extent that surface runoff becomes a rare occurrence. Where surface runoff does occur in the Horton manner the volume of water running off will increase downslope and distance from the watershed becomes a critical factor in how much erosion results. Precise volumes of water moving in this way will be hard to quantify in large drainage basins where there is unlikely to be uniformity in such factors as vegetation cover, land use or slope angle, the factors which affect infiltration and hence surface runoff.

Surface runoff *will* occur when the ground is totally saturated but under such conditions moisture is more likely to move into stream channels *through* the ground. Even following prolonged dry periods water will move at depth into stream channels. The practical experiments of Whipkey in 1965 and

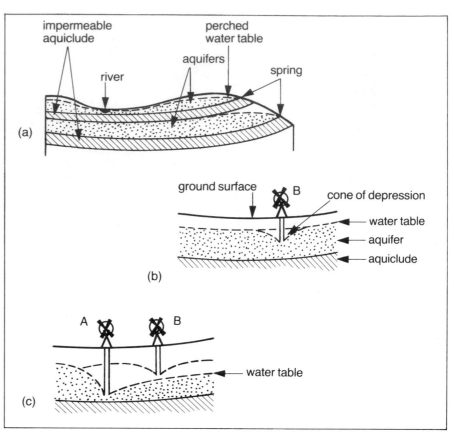

Fig 3.5 (a) aquifers and aquicludes (b) cone of depression (c) well capture: farmer A digs a deeper well and perhaps extracts water at a greater rate than B. A's cone of depression then captures water formerly going to B, whose well dries up

Hewlett and Hibbert in 1966 illustrate this process.

Hewlett and Hibbert simulated a storm of 5.1 cm/hr for 2 hours on a soil tank inclined at 16° to the horizontal and then measured the volume and duration of the discharge from varying depths within the soil. Their results are illustrated in figure 3.6. They show convincingly that water continues to move through soil long after the 'rainstorm' ended. Even though surface runoff may have occurred briefly on surface litter during the storm, in the Horton manner, it is the water moving through the soil at depth that contributes the majority of water to the theoretical stream channel at the lower end of the soil tank. Moreover it is the water at greatest depth that discharges for the longest period. In a sense therefore the throughflow theory can be thought of as a series of inputs of precipitation infiltrating the soil, each one pushing a previous water input into the stream at the foot of the slope (see fig 3.7).

One can conclude that both models

have a place in connecting rainfall and infiltration to stream discharges. The Horton model is particularly useful in arid and semi-arid areas, but it is likely that throughflow is the most important model in the humid tropical and moist temperate zones of the world.

Whether water runs off or through the ground surface has a considerable bearing on the soluble and sediment load of rivers. Figure 3.8 shows a simple analysis of the factors affecting the chemical composition of soil moisture and the way in which leaching of soil moisture can influence the quality of water moving into drainage systems and into underlying solid geology where chemical activity continues to take place.

Figure 3.9 shows how the chemical concentration of stream discharges is related to levels of flood discharge, while figure 3.10 shows how the soluble load of rivers varies following a series of floods in quick succession.

Sediment yield on the other hand depends on the macro-scale influences of climate, which determines type and

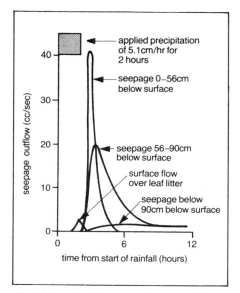

Fig 3.6 *Soil throughflow variation with depth*

rate of weathering, geology and, on a smaller scale, raindrop impact, sub-surface wash and surface runoff.

The start of stream channels

The way in which rainfall runs off or through the soil surface affects the way in which channel flow begins. It is useful to examine the theories as they have evolved through time. Horton's theory suggests that water infiltrates the soil to feed the water table which, when it intersects the slope surface, causes channel flow to begin. However, above this point on the slope 'rills' may form. They occur when precipitation exceeds infiltration so that water runs off on the surface, firstly as a sheet and then in a more concentrated linear form. The point on the slope at which these rills develop varies enormously (fig 3.11), and the distance from the highest point of the divide between drainage basins that the surface runoff occurs is called the x_c distance. Unvegetated slopes will have a much shorter x_c distance than well-covered slopes (fig 3.11(a)).

As throughflow became more widely accepted so there developed a second theory for the development of channel flow (fig 3.11(b)). This theory emphasises the continuing nature of water movement into and through the soil and suggests that channel flow on a slope may be ephemeral (the equivalent of rills), intermittent (like the winter-bournes) and permanent, depending on the amount of water in the soil and geology.

Eventually this theory too was seen to have limitations particularly in that

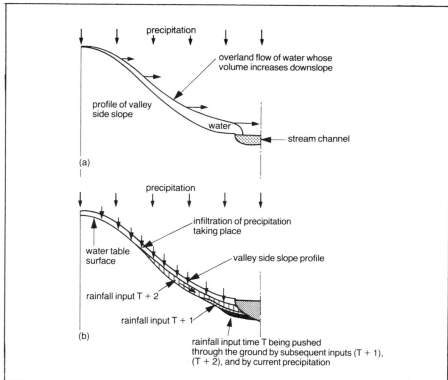

Fig 3.7 *(a) the Horton runoff model   (b) the throughflow model*

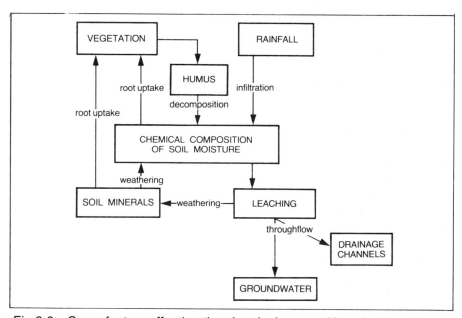

Fig 3.8 *Some factors affecting the chemical composition of soil moisture*

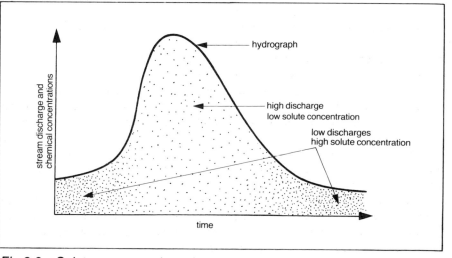

Fig 3.9 *Solute concentration related to levels of discharge*

it took into account only two dimensions, length and depth, whereas breadth should also be considered. The result is a three dimensional model of how streams begin, what has been called the expanding contributary area (fig 3.11(c)). In this model the importance of throughflow is realised but to it is added the importance of saturated flow from the valley bottoms or slope bottoms. Thus in a sense the model is a combination of the Horton and throughflow models.

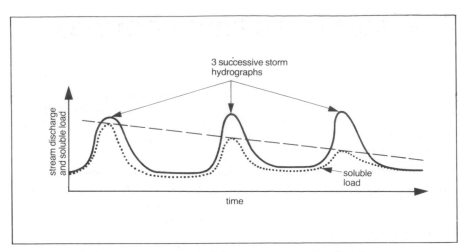

Fig 3.10   Solute concentrations following successive floods

EXERCISES

3.1 Study figure 3.12
(i) Why does the infiltration rate decrease through time?
(ii) How long after the start of precipitation does water start to run off the ground surface? How long does water continue to do so?
(iii) What is the total amount of precipitation that falls? What proportions of the precipitation enter the stream channel as surface runoff and throughflow?
(iv) How would your answer differ under the following conditions:
(a) twice the amount of rainfall;
(b) increased vegetation cover;
(c) the same precipitation on a less steep slope.

3.2 Measuring infiltration rates. Cut both ends from a canned fruit tin approximately 20cm in diameter. Draw circles inside the can at 5cm intervals. Take the can, with a water supply, to a hillslope near your home or school. Sink the can into the ground to a depth of 20cm. Fill the can with water to a depth of 15cm and record the time it takes the water level to fall 5cm. Repeat the experiment until the time taken is constant for at least three occasions. Express the infiltration in mm/hr.
   Repeat the experiment on different slope angles, soil types, vegetation types, and so on. Explain the variation of your results.

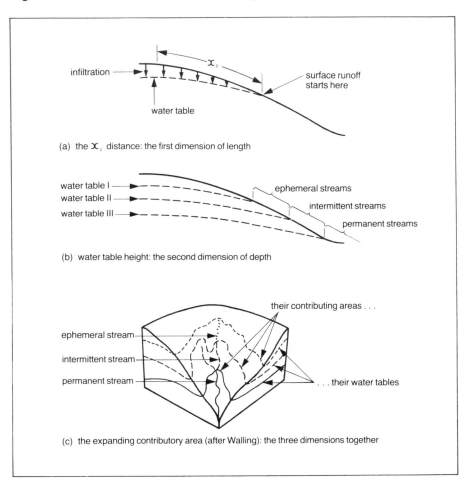

(a) the x_c distance: the first dimension of length

(b) water table height: the second dimension of depth

(c) the expanding contributory area (after Walling): the three dimensions together

Fig 3.11   The start of stream channels

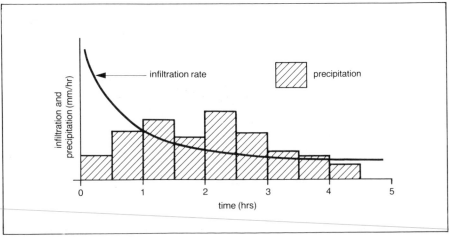

Fig 3.12

4 Water in stream channels

In previous pages we have seen how rain falls, infiltrates and moves into stream channels. It follows that we should consider how it behaves thereafter. The simple nomenclature in figure 4.1 will be useful to know.

Types of water flow

There are two basic types of water flow in stream channels, *laminar* and *turbulent* flow (fig 4.2).

Laminar flow occurs at low velocities within straight smooth river channels with layers of water shearing over each other dependent on the viscosity of the water. Because of the precise conditions needed for it to occur it is rarely found in stream channels (though it may exist during throughflow from soil and rocks into the stream channel).

A small deviation from the conditions needed for laminar flow will produce turbulent flow, a much more common type of flow in all stream channels. Important factors affecting it are channel roughness, water velocity, water depth and viscosity of the water (the latter often controlled by temperature). To distinguish between laminar and turbulent flow *Reynold's number* is used.

Reynold's number $\quad R = \rho\dfrac{VR}{\mu}$

ρ is water density
V is velocity of water
R is hydraulic radius
μ is water viscosity
small values of R indicate laminar flow
large values of R indicate turbulent flow

Two types of turbulent flow exist. The more common type is ordinary 'streaming flow' turbulence but at times 'shooting flow' turbulence may occur at much higher water speeds such as those found in rapids. The two

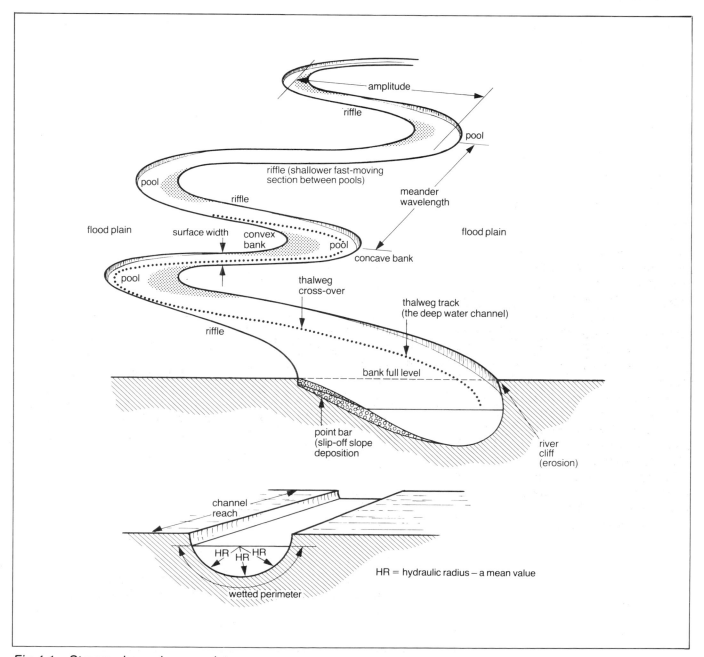

Fig 4.1   Stream channel nomenclature

types are distinguished by using the *Froude number* F.

$$F = \frac{V}{\sqrt{gD}}$$

V is mean velocity
g is the force of gravity
D is mean water depth
$F < 1$ indicates streaming flow
$F > 1$ indicates shooting flow, and water flowing in this way is usually very erosive because of its greater ability to transport debris

Whichever type of turbulent flow exists the velocity of the water in the stream channel will not be constant over the channel cross-section. Fastest flow usually occurs at the point in the cross-section furthest from the frictional drag of the channel sides in contact with the water (the wetted perimeter). It follows that the slowest rates of flow are found adjacent to the channel floor and sides. For those wishing to sample stream velocities the mean velocity of a stream is usually close to the value for 0.6 of the water depth.

Stream energy

The amount of work which water is capable of doing in a stream channel is a function of the amount of energy it possesses. The total energy available is mostly influenced by the velocity of the water, but the velocity is affected by gradient, the volume of water that is flowing, and the roughness of the channel. Approximately 95 per cent of a stream's energy is used in overcoming friction with the channel wetted perimeter, and internal friction within the water itself during turbulent flow. The importance of channel roughness in increasing frictional drag is obvious and its importance is recognised in *Manning's formula* which is used to measure empirically stream velocity (see table 4.1).

Manning's formula
$$\text{Velocity} = K \frac{R^{2/3} S^{1/2}}{n}$$

K is a constant, but varying according to units used
R is the hydraulic radius
S is slope of the water surface
'n' is Manning's coefficient for channel roughness, a figure which can vary considerably

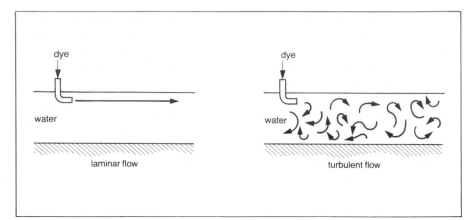

Fig 4.2   Laminar and turbulent flow

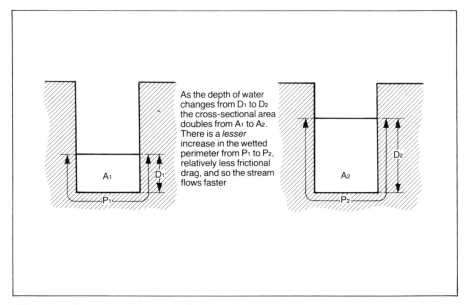

As the depth of water changes from D_1 to D_2 the cross-sectional area doubles from A_1 to A_2. There is a *lesser* increase in the wetted perimeter from P_1 to P_2, relatively less frictional drag, and so the stream flows faster

Fig 4.3   Stream energy and frictional drag

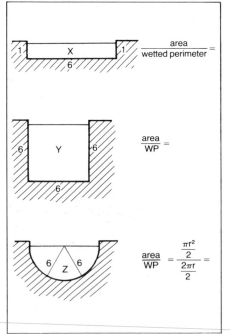

$$\frac{\text{area}}{\text{wetted perimeter}} =$$

$$\frac{\text{area}}{\text{WP}} =$$

$$\frac{\text{area}}{\text{WP}} = \frac{\frac{\pi r^2}{2}}{\frac{2\pi r}{2}} =$$

Fig 4.4   Stream energy and channel shapes

Table 4.1   Some values for Manning's coefficient 'n'.

Type of channel	Value of 'n'
Streams on plain	
Clean, straight, no deep pools	0.030
Sluggish reaches, weedy, deep pools	0.070
Mountain streams	
Bottom: cobbles, large boulders	0.050
Flood plains	
Pasture, short grass	0.030
Cultivated, no crop	0.030
Cultivated, mature crops	0.040

(Source: Gregory & Walling, *Drainage Basin Form and Process*, E.J. Arnold)

For those wanting to increase a stream's velocity, to get rid of water quickly for flood relief for example, the simplest methods are to straighten the stream's course, smooth its channel bed long profile, and to try to reduce the channel's wetted perimeter by making its cross profile as symmetrical and rounded as possible.

The effect of the wetted perimeter's length in relation to the volume of water flowing in determining the amount of energy a stream possesses (or the 'efficiency' of the stream channel) can be seen in figure 4.3 which shows that (a) an increase in depth of water greatly increases the amount of water flowing (indicated in this case by the change in channel cross section area from A_1 to A_2) but that (b) the increase in the wetted perimeter from P_1 to P_2 is proportionally much less and so there is less frictional drag on the water in the channel and so it flows faster.

EXERCISE

4.1 Look at figure 4.4. Calculate the efficiency of the stream channel shapes below by dividing the cross-sectional area by the wetted perimeter. The higher your number the more efficient the channel shape.

Measurement of river discharge

This is one of the most important characteristics of streams for hydrologists to know. A number of measurement techniques can be used and all express their results in cubic units per second.

The simplest method of measurement of discharge is velocity × area. The principle is simple. A channel cross profile is measured as accurately as possible, normal to the stream direction at a point free of water weeds and over 30cm deep, and drawn to scale in order to calculate its area. At regular intervals through the cross profile speed of flow is measured and a mean value obtained. The discharge is simply the product of the two figures. For precise figures of velocity a current meter should be used at 0.2 and 0.8 of the water depth and

this method gives the best results, but for those without this equipment, an orange and the second hand of a watch give surprisingly consistent results for small streams.

A more accurate method is to measure the volume of water passing a point in the stream channel in a given time. A suitable length of stream channel is chosen, a series of cross profiles of the channel are drawn and their mean area calculated. Mean cross profile area multiplied by the length of the channel chosen gives a figure for the volume of water contained therein. A mean value for the time taken for water to flow the length of the channel is then obtained. Thus, if 20m³ of water (the measured water volume) follows the length of the channel in 30 seconds (the measured mean time) then the discharge of the stream is 20/30m³/sec (cumecs). Figure 4.5 shows this method used to calculate the discharge of the River Yeo at Sherborne. The calculation is as follows.

Mean cross-sectional area =
$$\frac{3.4+2.4+3.6}{3} = \frac{9.4}{3} = 3.1m^2$$
Volume of water in channel reach =
$$3.1 \times 15 = 46.5m^3$$
Mean time taken for orange to
travel distance A–B = 24 secs

Therefore, as 46.5m of water pass through the channel reach in 24 secs, the river discharge is $\frac{46.5m^3/sec}{24}$

The discharge of the River Yeo was 1.9m³/sec

Using simple equipment such as knotted string (at 0.5m intervals), a metre rule (for plumbing depth), an orange, and the second hand of a watch, five groups of students working independently on the River Yeo recorded discharges varying no more than 0.15 cusecs from the figure given above.

EXERCISE

4.2 Using the method outlined in figure 4.5, calculate the discharge of a small stream near your home or school.

Hydrologists frequently want more regular readings of river discharges than the simple methods outlined above encourage, and so more permanent recording installations are set up, usually in the form of weirs or flumes. An excellent installation for measuring discharge on small streams is the sharp crested 90° V-notch weir (see fig 4.6). Using this method the height of water h passing through the notch is recorded and used in the equation

$$\text{Discharge} = 1.38 \, h^{2.5}$$

A variety of sharp crested weirs can be used but they all run the risk of being damaged by floods, and their readings can be disturbed by silt accumulating behind them. In such instances they are often replaced by broad-crested weirs or flumes (see fig 4.6) both of which are less susceptible

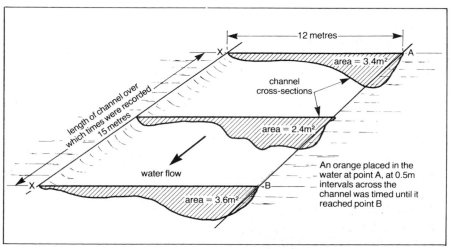

Fig 4.5 Measurement of the discharge of the River Yeo at Sherborne on 13 February 1979

to damage and in the case of the latter are self cleaning. Unfortunately they are more costly to construct. Their use is to hold back water and the depth of water upstream of the obstacle is measured. This has the advantage that it can be allied to a permanent water recording device similar to a cylindrical barograph.

A 'tell-at-a-glance' method is to firmly embed a calibrated pole into the river. This then acts as gauging post, there being, once records have been established for sufficient time, a good correlation between gauge post readings and river discharge (see fig 4.7).

It is also possible to measure discharge by dilution methods which are particularly valuable for extremely turbulent or vegetated channels unsuitable for current meters. Known concentrations of chemicals (often common salt) are added to a stream and then the concentrations downstream measured, usually by conductance of the water, but the techniques are too impractical for use at school level.

Hydrographs

Regular measurements of discharge enable hydrologists to construct hydrographs, i.e. patterns of river discharge through time. They may represent different time periods, and be for different purposes, but they all reflect the way in which a drainage basin responds to a given input of precipitation, the response being affected by the nature of the rainstorm and the physical characteristics of the drainage basin.

Annual hydrographs

These portray the discharge of a river throughout the year, the figures being the mean values of discharge records of several years. An excellent example is that of the Nile, capable of producing 'seven years of famine and seven years of plenty', as it enters Lake Nasser, to be regulated by the Aswan Dam. Figure 9.5 illustrates the relative contributions of the White Nile, Blue Nile, Atbara and Sobat rivers to the total discharge of the Nile, and the time of year when the contributions to the famous Nile flood are made. One can detect the White Nile, intercepted by the great Sudd swamp of the Sudan, forming the 'base flow' of the Nile's discharge while the Ethiopian monsoon-fed Blue Nile provides the bulk of the annual silt-laden flood once so important for Egypt's economy.

Storm hydrographs

These show the discharge pattern of a river following a single storm. A classical storm hydrograph is shown in figure 4.8 and it shows how, prior to rain falling, river discharge is supplied by throughflow from previous storms stored at depth in soil and rock. During and following the rainstorm the river discharge rises, initially due to direct precipitation into the stream channel (not subject to a time-lag and therefore fast acting), then perhaps some surface runoff over leaf litter, and then followed by throughflow. These components all account for the rising limb of the hydrograph but the recession curve following the flood peak is entirely the

Fig 4.6   (a) sharp-crested V-notch weir

(b) broad-crested flat V weir

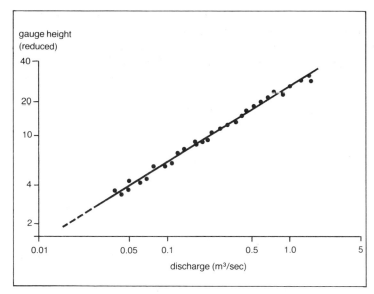

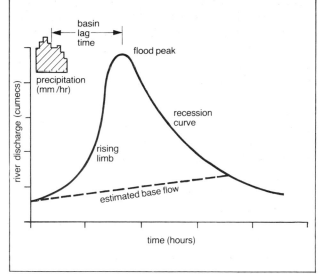

Fig 4.7   A stage-discharge graph – useful for quick recording of river flows. The y-axis has been reduced so that a gauge reading of zero represents nil discharge

Fig 4.8   The storm hydrograph

result of throughflow. As its name suggests 'basin lag time' is the time taken for the drainage basin to respond to the rainstorm and produce a flood peak. It is very hard to assess the precise contribution of ground water 'base flow' to the hydrograph and most authorities accept that it gradually increases during the period of flood and slowly declines thereafter.

A little thought will make it clear that different character rainstorms will produce different storm hydrographs, and to make the job of predicting discharges from given drainage basins more accurate, *unit hydrographs* are used.

Unit hydrographs

These hydrographs were first devised in an attempt to simplify the relationship between precipitation and resulting stream discharge. Unit hydrographs assume that a given storm input into a drainage basin (usual inputs used are one inch or one centimetre of rainfall covering the entire drainage basin in time T) will produce a constant runoff pattern in the stream. Hence 1cm of rainfall in one hour will produce runoff pattern A in figure 4.9(a) and if different T values are used e.g. 1cm of precipitation in 5 hours, then the unit hydrograph will be different (runoff pattern B).

To derive a unit hydrograph information is needed about precipitation and runoff following a given storm. If during a storm 1.5cm of rain fell then the plotted hydrograph would be reduced to the equivalent of 1cm of precipitation by reducing the ordinates by one-third as in figure 4.9(b).

The wide variety of factors which can influence the form of a hydrograph, which are illustrated below, make it clear that a unit hydrograph cannot be totally reliable, but given an awareness of this fact it can be of immense value in predicting flood discharges, and therefore in reducing the cost of flood damage.

Factors affecting hydrograph forms

The large variety of factors that can affect a hydrograph fall into two categories. First, the *permanent* characteristics of the drainage basin such as its size or drainage density, and second, transient, or variable characteristics such as the amount of precipitation, type of land use and so on. These two sets of factors have been categorised as shown in figure 4.10. As can be seen the range of factors is considerable (some will be dealt with in future chapters) and the larger the drainage basin the less chance there is of their being uniform throughout the catchment area. This means that the chances of apportioning values to each of the factors becomes more difficult as drainage basin size increases and so predicting discharges of large rivers precisely is a demanding exercise. A certain measure of success can be

obtained if the drainage basin has a dense cover of recording stations but often these are less frequent than recording points for weather phenomena and so it is the latter that provide much of the information for the hydrologist's predictions of floods.

Urban hydrographs

One of the most interesting modifications of hydrographs takes place following the growth of urban areas within catchments. Urban surfaces are mostly tarmac or concrete and largely impermeable as a result. Natural infiltration is therefore negligible and surface runoff correspondingly rapid and on a large scale. To avoid the inconvenience of surface water in towns artificial drainage systems are installed which, perhaps, are a man-made equivalent to the macropores mentioned on p. 12. The result is rapid entry of surface water into subsurface storm drains or, in the case of old towns, into the sewerage network which it was intended to flush clean. These artificial drainage systems lead eventually into natural channels where the water discharged can have remarkable effects on the river hydrograph. Figure 4.11 shows the 'before and after' effects of urbanising a catchment area. Flood peaks are often increased several hundred percent and, in the absence of gentle throughflow and soil storage, the basin time-lag and recession curve are both reduced. Floods occur more regularly and the more sudden discharge into the natural channels would drastically modify their character, usually scouring the channel bed because of a reduced supply of debris. The character of the water being discharged would be different too; it would be marginally warmer and contain different dissolved and sediment loads from neighbouring rural streams.

All these effects of urbanisation come to a head *during* the building period when natural surfaces are destroyed, compacted, and not yet replaced by man-made substitutes. Work in Exeter by Gregory and Walling quotes erosion rates 200 per cent greater than normal during a building period in part of the city.

River loads

Debris is transported by a river using the 3–5 per cent of its energy not needed to overcome friction, and the total amount carried will be a function of the river's discharge. The river's load can be classified into the following groups.

Flotation load
This is a relatively unimportant type of river load consisting of organic matter such as leaves, branches, or even trees carried mostly during floods. It can be an important type of load if the debris accumulates to form temporary dams which divert the course of a stream, or which suddenly give way to produce water surges downstream. Much of the damage of the famous Lynmouth flood of 1952 was thought to be due to the collapse of temporary barricades allowing surge waves to move downstream.

Soluble load
The accumulation of lime inside domestic kettles is the most familiar indication of the dissolved mineral content of water, and there will be no stream channel anywhere that contains water without some degree of minerals in solution. On a large scale the amount of the soluble load will be the result of climate which influences the rate of chemical weathering of rocks which provide many of the dissolved minerals, and also the rates of chemical reactions when minerals are dissolved. Geology,

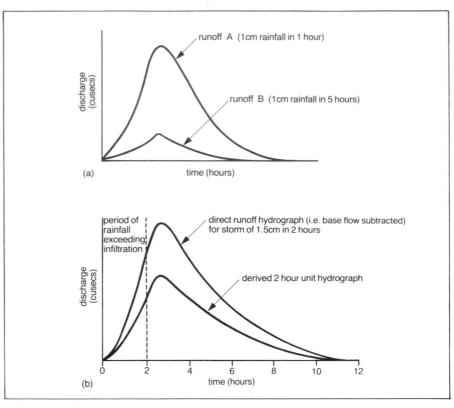

Fig 4.9   The unit hydrograph   (a) different T values
(b) derivation of unit hydrograph

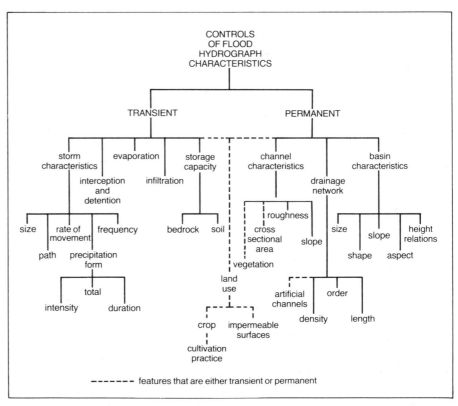

Fig 4.10   Factors affecting the characteristics of the flood hydrograph

too, will be important; some rocks, notably limestones, are more soluble than others, while the decay of vegetation provides much of the acid content of water and so determines its power as a solvent. A consideration of the origins of the soluble load appears above in the section on ground water (p. 15).

The amount of the soluble load varies

enormously from one stream to another and streams that owe their origin to ground water more than surface runoff sources tend to have a higher soluble load, e.g. the English chalk streams whose dissolved minerals sustain both prolific plant and fish growth. Variations in the chemical content of the soluble load are at their maximum between small drainage basins because these might have local peculiarities which could encourage high or low values of one mineral or another. For rivers draining large catchment areas there is considerable consistency in their soluble loads and bicarbonate, chloride and sulphate of calcium and sodium make up more or less 90 per cent of the soluble load.

The soluble load of streams can be measured by passing an electric current through the water to measure its conductance which increases as the total dissolved load increases. The same technique can be used to measure concentrations of individual salts in solution (see fig 4.12).

Sediment load

Table 4.2 shows the San Juan river (Utah) carrying 97 per cent of its total load as sedimentary load. Such a figure is by no means exceptional but does reflect the nature of the local geology and prevailing weathering processes which combine to determine the nature and volume of the debris supplied to the river. The sedimentary load is usually sub-classified into bed load and suspended load though there is a grey area between the two of temporarily suspended debris.

EXERCISE

4.3 How would you account for:
(i) the levels of suspended sediment in table 4.2(a)?
(ii) the range of suspended sediment yields in table 4.2(b)?
(iii) the variation in suspended and soluble loads in table 4.2(c)?
You will find it useful to consult atlas data on climate, soil types, vegetation and land use.

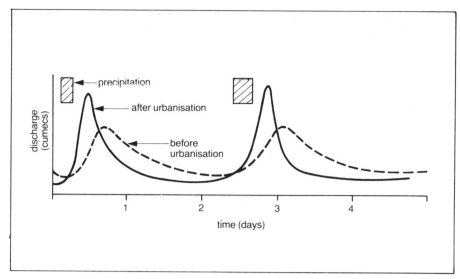

Fig 4.11   The effect of urbanisation on the hydrograph

Table 4.2 Some examples of sediment yield in major world rivers

a) Some maximum recorded suspended sediment yields

River	Country	Average suspended sediment yield p.a. (tonnes/km²)
Ching	China	8040
Lo	China	7922
Waipaoa	New Zealand	6983
Tjatjabon	Java	6250

b) Suspended sediment yields in major drainage basins

River	Country	Yield (tonnes/km²)
Ganges	Bangladesh	1568
Colorado	USA (Arizona)	424
Mississippi	USA (Louisiana)	109
Amazon	Brazil (Belem)	67
Nile	Egypt	39
Rhine	Netherlands	3.5

c) The relative importance of sediment and soluble load in selected rivers

River	Country	Sediment (% of total load)	Solute (% of total load)
San Juan	USA	97	3
Colorado	USA	94	6
Tyne	UK	65	35
Don	USSR	45	55
Pilica	Poland	7	93

The proportion of bed load is critical as a channel forming factor and as its name suggests it is the material, usually coarse textured, that moves downstream in contact with the stream bed. Whether a particle on the stream bed moves will depend on its volume, density and shape (round stones move more easily than flat ones), and the force exerted on it by the moving water, often called the *critical tractive force*. It can be shown experimentally that there is a strong positive correlation between the weight of particles and the velocity of water needed to set them in motion. It can also be shown that once set in motion large bed load particles move faster than small ones which are more affected by turbulence or upstream eddies, and which also tend to fill in the spaces between the larger particles thereby coming to rest.

Bed load is trundled along the stream bed but some debris may be set into temporary suspension by random hydrodynamic forces which have the greatest effect on the smaller elements of the bed load. As these disturbed particles come into contact again with the stream bed downstream they may disturb other particles into motion: such an intermittent downstream movement of bouncing particles is called *saltation*.

Two useful terms for use in connection with bed load transport are *competence* and *capacity*. The former refers to the largest particle that can be moved at a given discharge, while the latter refers to the total bed load of a certain calibre (size) that the river is capable of moving. Obviously neither figure is a constant and both rise and fall in proportion to variations in discharge and come to a peak at times of flood.

The suspended portion of the sediment load consists of particles fine enough to be kept in suspension by the turbulence of the water, though it must be remembered that particles kept in suspension at one level of discharge may not be carried thus if the stream's discharge decreases. The highest concentrations of suspended load are near the stream bed, the lowest amount in the part of the stream channel where turbulence is least, usually near the surface. There is, as with the critical tractive force needed to move bed load, a sensitive interaction between the size of particle that can be kept in suspension and the discharge and turbulence of the stream flow. This can be illustrated using a glass tank, sediment of various sizes and a time-piece. A graph plotting the time taken for different sized particles to reach the floor of the tank produces consistent results, but when subjected to water turbulence the figures change considerably (see fig 4.13).

It is possible to produce a diagram based on Hjulstrom's work which illustrates the links between water velocity and size of particles that are moved or deposited (see fig 4.14). Large particles usually fall faster than small ones and so will be the first to be deposited when discharge falls and the last to be moved as discharge increases. This relationship is more constant for spherical particles than debris of assorted shapes (flat stones are usually slow-sinking) and the density of the particles will also affect their behaviour.

At the risk of labouring the point it is important to appreciate that the movement and deposition of debris within

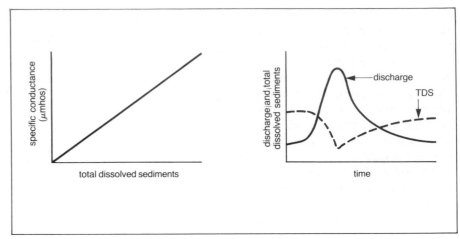

Fig 4.12   Generalised relationships between soluble load and conductance

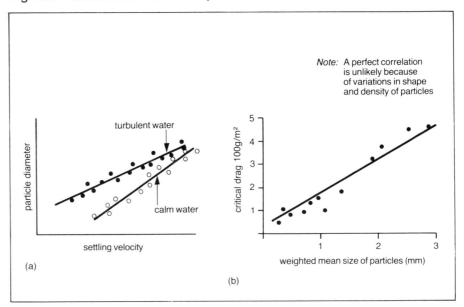

Fig 4.13   Particle size and stream energy
(a) settling velocity of particles
(b) critical tractive force needed to set particles in motion

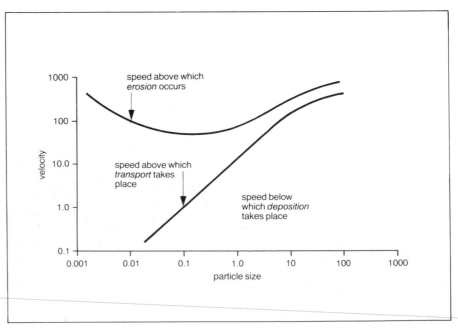

Fig 4.14   The Hjulstrom curves

26

the stream channel is a dynamic process. As discharge increases or decreases there is a corresponding increase or decrease in the amount and type of debris transported. This manifests itself in the changing forms within the stream channel. Anyone who knows a small stream intimately or who has the time to make detailed observations of one over a period of time, will be well aware that the shape of the river channel is never constant and is forever changing in response to the varying volumes of water and debris passing along it, and that the most dramatic changes take place following high water levels. For example, a sandy stream bed will, at low water levels, often be formed into a series of small ripples or dunes. At higher water levels the dune pattern changes as the dunes become fewer but larger in size to match the different flow patterns. Individual pools, too, can change dramatically and figure 4.15 shows a stretch of the River Tywi in west Wales where, after a wet winter and several high floods the stretch of water had changed very considerably.

Measurement

Methods of measuring the sediment load are well documented and equipment, such as the cross channel bed conveyor belts used to extract bed load in some of the North American rivers can be large and expensive. However, on a smaller scale one can also obtain useful results by more simple methods.

Bed load can be fun to measure. The simplest method in a small stream is to sink containers (large biscuit tins will do) into the channel bed flush with the stream bed, into which debris falls. They need to be regularly emptied of debris to be effective. A refinement of this method is a fine chicken wire 'basket' to fit the stream channel: debris moving along the stream bed will enter the basket but will not leave it. Both these devices can be difficult to secure in large or rocky streams. Bed load particles can be sieved to determine their size and, if the particles moved are large enough it is possible to measure pebble roundness using the Cailleux formula. Interesting data can thus be obtained of the mean size and standard deviation of debris transported at given discharge or at different

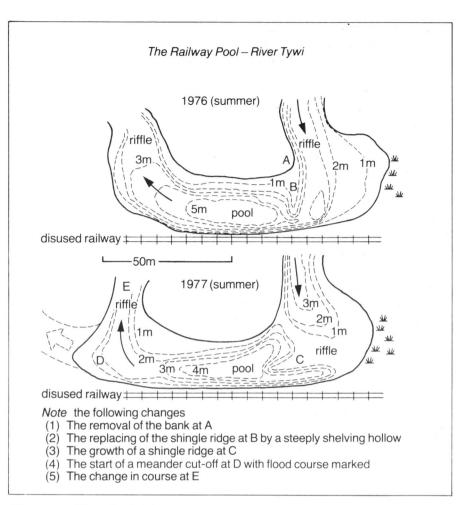

Fig 4.15   River Tywi, West Wales

Fig 4.16   A DH 59 suspended sediment sampler

locations along the stream's course.

Suspended sediment can be measured professionally using a sediment sampler, which is a fish-like device whose fins orientate it in the direction of stream flow and which contains a storage jar inside (see fig 4.16). On small streams milk bottles attached to a vertical post can be used (see fig 4.17) but this method makes samples close to the stream bed difficult to obtain. The same basic principles apply both to the sophisticated and the simple methods, namely to take samples at varying points over the channel cross-section and to try to ensure that there is minimal disturbance of flow patterns by the bottle or the sampler! Water samples containing the suspended sediment can be filtered and the amount of suspended load expressed in milligrams per litre of water. Fine sieves can also be used to get indications of the different sized particles being transported in suspension, but for tiny streams the sieves are not really fine enough for the calibre of material in suspension and instead, a centrifuge like those found in most school chemistry laboratories should be used to isolate the suspended material.

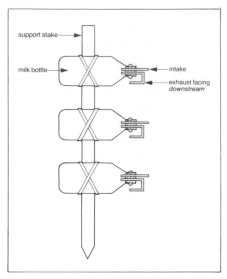

Fig 4.17   A home-made suspended sediment sampler

5 Stream channel processes and patterns

Previous chapters have referred to the way in which water moves in stream channels and to the ways in which debris is transported. The movement of debris-laden water has, in turn, an effect on the characteristics of the channel in which it flows and, on a larger scale, on the landforms within the entire drainage basin. The study of these effects constitutes the field of physical hydrology, and will be examined in the following pages, prior to moving on to the more anthropocentric aspects of the subject, namely applied hydrology – the way in which water is used by man.

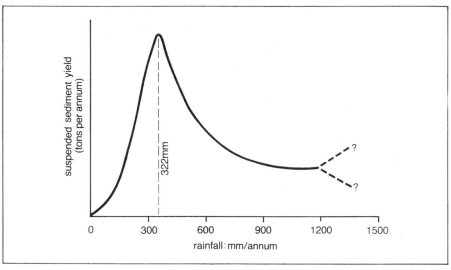

Fig 5.1 Sediment yield and annual precipitation

Erosion and deposition by water

Erosion and deposition within the stream channel determine the channel landforms, and the debris being carried by the water acts as the tools of erosion as well as constituting the material that is deposited.

Erosion within stream channels takes place in three main ways: *corrosion, corrasion* and *hydraulic action*. Corrosion is the chemical reaction between water in the stream channel and the rock composing the channel's wetted perimeter. The higher the water temperature the more likely it is that corrosion will occur, but the nature of the rock is important too. For example, limestones are much more susceptible to this form of erosion than any other rock in temperate latitudes.

Corrasion is the wearing away of the stream channel by the debris carried by the stream, particularly the bed load. Not only can the stream channel itself be worn smooth by this process, but the debris load can also be worn down and rounded by grinding against itself, a process called *attrition*.

Hydraulic action is the force of the water unarmed with debris. It is most effective in removing unconsolidated material from river channels and so is most important in alluvial stream channels. A more specific type of hydraulic action is cavitation, a process associated with high water velocities such as those found at rapids or waterfalls when air bubbles in the water implode, setting up shock waves which are readily transmitted through the water against the banks of the channel.

As a result of these three processes the landforms within the stream channel, and the course of the stream itself, may change. The effectiveness of erosive processes is measured by the amount of sediment they produce for the river to carry away, and effectiveness is largely determined by prevailing climatic conditions which provide the water, and the geology of the area which provides the resistance to the eroding processes.

An attempt to relate erosion to precipitation is shown in figure 5.1, based on work by Langbein and Schumm in the mid-west of the United States. This suggests maximum sediment yield in areas experiencing 300–325mm of rainfall per annum. Below this figure there would be insufficient rainfall to be a really effective eroding agent, while above 325mm the extra rainfall would encourage an increasingly effective vegetation cover which would reduce the potential erosive powers of higher rainfall. In other words at the 300–325mm mark there would be enough rainfall to encourage runoff in excess of infiltration, but not enough to produce protective vegetation. When rainfall totals exceed 1000mm it becomes a subject of much modern debate whether the increased rainfall overcomes the effects of vegetation or not. While it is unlikely that these findings would be applicable to all parts of the world they do suggest that the answer is not a simple matter of reconciling only two variables, and other factors come into play, not least types of vegetation and slope angle.

It was shown in the previous chapter that as discharge decreases, less debris can be carried and when this occurs deposition takes place. Several factors can reduce the carrying capacity of water causing deposition to take place. Some of the more important causes are listed below.

(1) A reduction in water velocity due to a change in the stream's gradient, perhaps a result of a geological change in the river's course.

(2) A reduced flow due to abstraction of water by man. Evaporation downstream can have the same effect in exotic rivers such as the Nile or the Colorado.

(3) Geological changes which can change the chemical nature of water in the stream. These in turn encourage dense vegetation growth in the stream channel which slows the water movement.

(4) Additions of extra debris to the stream, perhaps as a result of man's misuse of adjacent farm land or due to debris coming from a very steep and active tributary, which cannot be carried by the less energetic main stream.

These causes of deposition may operate singly or in varying combinations, but the single most important factor is the reduction in the speed at which water flows. Once this has happened then deposition is inevitable, obeying the simple dynamics outlined on pp. 25–27.

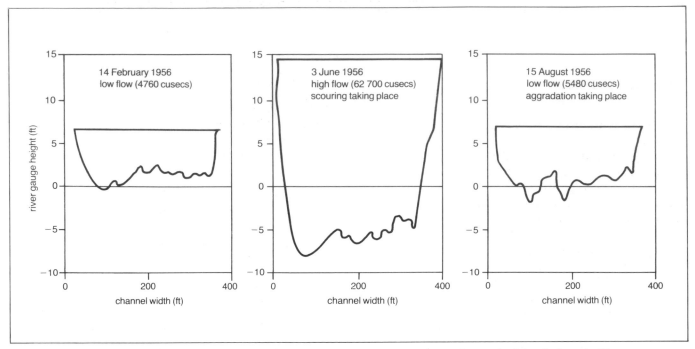

Fig 5.2   Aggradation and degradation in stream channels. Changes in the channel of the Colorado River at Lees Ferry

Other changes in a stream's ability to transport or deposit debris take place in a much longer time scale. For example, mountain-building activities may lower or lift up a landscape; changes in sea level can produce different base levels to which streams erode; prolonged denudation can progressively lower a landscape. All these changes can increase or decrease a stream's ability to erode, transport, or deposit material, primarily by altering the overall gradient from source to mouth of the stream.

It is, however, wise to point out that changes in carrying capacity can have both local effects, that is, at a *point* in the stream's course, and effects on a much larger scale throughout the *length* of the stream's course. The controls on both sorts of change are likely to be different.

It is very common for both erosion and deposition to take place at the same point of a stream's channel but at different times, particularly in those climatic regions which generate strongly seasonal flows. The result is that at high discharges erosion processes will dominate and the stream bed will be scoured and deepened; subsequently, at a lower discharge, the same part of the channel may 'aggrade' and become shallower because of the debris deposited in it (fig 5.2). Furthermore it is possible for both erosion and deposition to be occurring simultaneously in the same part of the stream channel. The most obvious example of this occurs on curved reaches of rivers

Fig 5.3   Meanders and point bars: the River North Tyne

where the outer bank of the river has the deepest water at its base and is being eroded into a typically undercut form while, simultaneously, the inner bank of the bend may be extending out into the channel by means of a depositionary point bar (fig 5.3). The point bar accumulates debris, particularly at high discharges, which then becomes colonised by a succession of plants, becomes stable, and so fixes a new course for the stream. Erosion and

deposition are able to operate simultaneously in this instance because of the helicoidal (corkscrewing) motion of water as it flows round bends (fig 5.4). On the Colorado river the large scale controlling of water has led to such a regular discharge that the supply of debris for maintaining the existing point bars has been reduced, so much so that the bars are being eroded as the Colorado seeks debris to carry to use the excess energy which it possesses.

Channel geometry

Even finer adjustments of response to changes in process can be seen in the geometry of stream channel sections.

(1) There is a good positive correlation between the discharge of a river and the speed at which water flows in the channel, the depth of the channel, and the width of the channel. To illustrate this, data is usually obtained along a river's course from source to mouth and the resulting wide range in the value of the data is overcome by expressing the relationships using a logarithmic scale (fig 5.5). Discharge has been interpreted here as the discharge at bankful level because this is the discharge most likely to be responsible for determining the form of the channel.

(2) The same sort of relationships between discharge, speed, depth and width can be observed at a *single* point along the stream's course. Though the interrelationships are very constant for the same point on the same stream a similar increase in discharge on two streams flowing over different substrata will produce different channel form responses. For example, an increase in discharge in a stream flowing over cohesive bedrock will tend to increase the depth but not the width of the channel. A stream flowing over unconsolidated material on the other hand will increase only slightly in depth but considerably in channel width.

(3) The influence of debris type on channel form, hinted at above, can be illustrated by comparing the width/depth ratio with the proportion of the debris load composed of silt or clay. As the proportion of fine silt or clay increases in the stream's load, the width/depth ratio decreases.

If one combines these three observations, then it becomes obvious that the form of a stream channel, its channel geometry, at any point in the stream's course is the result of a combination of stream discharge, the volume and nature of the debris being carried, and the type of geology over which the river is flowing. The effect of this interrelationship is made clearer when considering channel patterns.

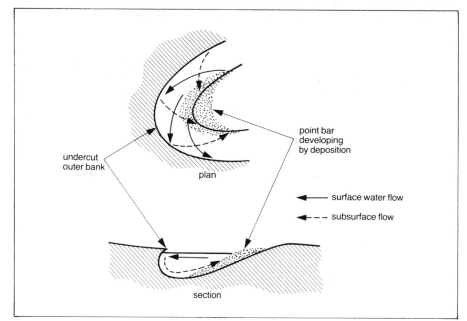

Fig 5.4   Helicoidal flow of water in meandering stream channels

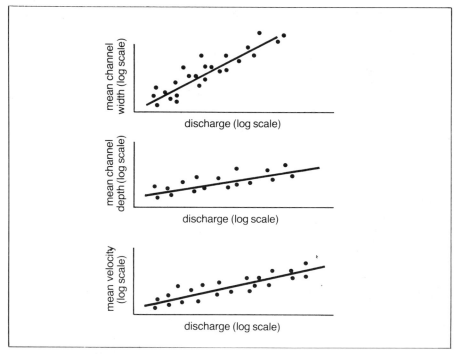

Fig 5.5   Relationships between changing mean annual discharge downstream and mean channel width, mean depth and mean velocity

Channel patterns

Seen from above few stream courses are constant over any length, nor are any two patterns exactly alike. For example, perfectly straight stretches of natural channels are extremely rare and even where they occur they rarely exist for distances much in excess of ten times the channel width. Needless to say, the frequent deviations from the straight line can be slight or extensive and the sinuosity of stream channel patterns is correspondingly varied. To overcome the problems of vagueness associated with descriptive methods of classifying channel patterns (e.g. where does a 'slightly meandering' channel become a 'moderately meandering' one?) sinuosity is expressed in numerical terms which are unambiguous. To obtain a value for sinuosity the length of stream channel is expressed as a ratio of the valley length and values of sinuosity can be near 1.0 for virtually straight channels (low sinuosity) and in excess of 4.0 for very sinuous streams (see fig 5.6). While accepting the difficulties of too coarse a classification, stream channel patterns can be grouped into three types: straight (sinuosity <1.5), meandering (sinuosity >1.5) and braided.

EXERCISES

5.1 Calculate the sinuosity index for the two streams (Y and Z) shown in figure 5.6

5.2 Trace from a 1:25 000 Ordnance Survey map portions of river courses and work out their sinuosity indices by the method shown above. For further work, correlate sinuosity indices with the gradients of the sections you have chosen.

Is there a relationship between sinuosity and gradient? If not, why not?

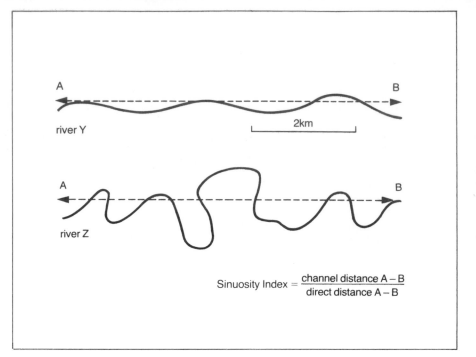

river Y

2km

river Z

$$\text{Sinuosity Index} = \frac{\text{channel distance A} - \text{B}}{\text{direct distance A} - \text{B}}$$

Fig 5.6   Calculation of sinuosity

We have seen that straight channels are unusual and even when they occur the channel cross-section is rarely constant and the deepest water, the thalweg, moves from one side of the channel to another (fig 4.1). Well developed pools and riffles are not common in straight channels except those where the debris carried ranges widely in calibre. One further unusual feature of some straight channels is that they tend to have a central ridge of desposited material, clearly observable on the channel cross-profile, possibly due to the water flow pattern which consists of two 'tubes' of faster moving water parallel to the stream banks converging in the centre of the river (see fig 5.7).

Meandering channels are defined as having a sinuosity in excess of 1.5, and meandering is a characteristic of most streams (fig 5.8). Because they are the most common form of channel pattern, a lot of research has been done on meandering streams, not least by simulating meandering under laboratory conditions, and it has been found that a number of complex interrelationships exist which are as true for tiny streams as for the Mississippi. Below are some of the relationships which appear to exist between meander parameters, and the terminology may be better understood by reference to figure 4.1.

(1) Early theories seeking to explain meanders by relating them to the rotation of the earth have been abandoned.

(2) Meander wavelength is normally between six and ten times the channel width.

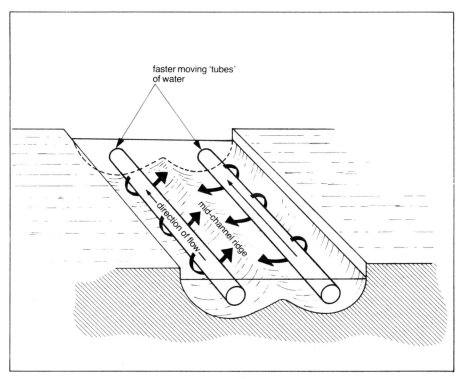

faster moving 'tubes' of water

direction of flow

mid-channel ridge

Fig 5.7   Theoretical flow of water in straight channels, creating a mid-channel deposition ridge

(3) The width of valley floor over which meanders swing (the meander belt) is normally between fourteen and twenty times the channel width.

(4) There is a good correlation between meander length and radius of curvature.

(5) The riffles associated with the thalweg crossover occur along the channel at approximately six times the channel width.

(6) Sinuosity of meanders increases as the depth of the meandering channel increases in relation to the width of the channel.

(7) Meandering becomes more pro-nounced as bed material becomes more varied in calibre, and meander wave-length increases in streams which carry coarse debris.

(8) Relationships of meandering to discharge are more obscure. It has been suggested (by Leopold, Wolman and Miller) that meander length is empiri-cally related to the square root of the dominant discharge but probably the relationship is not that simple and channel gradient is often inserted into the equation which reveals meandering more likely to occur at lower slope angles.

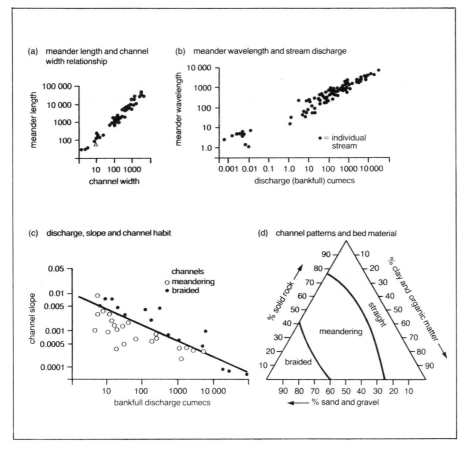

(a) meander length and channel width relationship

(b) meander wavelength and stream discharge

(c) discharge, slope and channel habit

channels
o meandering
• braided

(d) channel patterns and bed material

Fig 5.8   Meandering

Clearly meanders are not simple things (fig 5.9) and despite the observations above, some matters remain uncertain. For example, meanders can occur on ice or solid rock as well as the more 'normal' bed material. This raises the question of whether they are initiated in the same way in each environment. An explanation may lie in discordant drainage causes. Meanders may have developed on one type of surface which, as it became eroded away, superimposed the meander pattern on very different surfaces beneath. Alternatively, meanders may have been part of a drainage pattern which maintained its shape as the land became uplifted around it into a different configuration; in other words antecedent drainage. One may be tempted to accept these explanations for meanders such as those of the Wye at Chepstow, or the Rheidol inland of Aberystwyth, but they are somehow less palatable for the hugely-incised meanders of the Colorado in its canyon in the arid south-west of the United States.

Perhaps the best proposed explanation for meander development is that it occurs in conditions where channel slope, discharge and bed load combine to create a situation where meandering is the only way that the stream can use up the energy it possesses equally through the channel reach. We are thus moving towards the idea of *drainage basin equilibrium* (pp. 34–38).

A braided stream channel is one in which the body of water is divided and separated into a number of anastomosing (rejoining) channels by islands of debris (see fig 5.10). Braiding is associated with streams whose discharge fluctuates considerably and which are supplied with large amounts of mixed calibre debris. It is this debris which is picked up or deposited within the banks of the channel by the river at varying discharges. Thus, while the channel at peak discharge may contain water from one bank to another, transporting large amounts of debris, often, interestingly, in a straight course, the same channel at low discharges may contain many smaller streams flowing along complicated courses between banks of freshly deposited gravel. The mechanisms of braiding are reasonably straightforward. As discharge falls coarse debris can no longer be transported and is deposited to form the resting place for further deposits. The accumulating deposits may cause the stream to diverge to either side where the concentrated flow causes erosion of the channels to leave the original deposit exposed as an island or bar. Sometimes these midstream bars become colonised by vegetation and become semi-permanent, but more often than not they last until the next peak discharge.

Braided streams therefore have notably unstable channel forms and some spectacular examples of braiding can be seen where streams leave glaciers. The amount of ice-melt varies from one season to another and hence the discharge varies considerably; the glacial moraine provides abundant quantities of mixed debris and the two factors combine to create ideal conditions for braiding. Leopold, Wolman and Miller quote the Khosi river emerging from the glaciers to the south of Mount Everest changing its channel by as much as 19km in one year, and in the last 200 years it has moved over 112km to the west across the huge debris cone that it has deposited.

Other conditions which can promote braiding include seasonal monsoonal rainfall which can cause the channels of monsoon-fed rivers to change by over a hundred metres a *day*. Similar high-intensity precipitation can also be a feature of arid and semi-arid areas where, if debris is available as it often is in the absence of protective vegetation, stable braided river patterns exist until the next deluge changes them. In each of these examples of the causes of braiding, their effectiveness is enhanced if the varying discharge is accompanied by a change in the river's course from a high to a low relief area. This factor of change in slope angle is a feature of meandering too, and a relationship can be expressed graphically (see fig 5.8). Braiding, despite being a less frequent type of channel pattern than meandering, is probably better understood, though quantifying cause and effect has not been done on any large scale and a braiding index is one of the few mathematical analyses.

Brice's braiding index

$$BI = \frac{2l}{m}$$

where l = sum of the length of islands and bars in a given channel reach

m = length of the reach measured between points midway between the banks

BI > 1·5 is reckoned to indicate braiding

Drainage networks

A brief perusal of the standard geomorphological texts of yesteryear will reveal all manner of attempts to recognise and define drainage patterns. The definitions which were attempted were based on descriptive terminology and the classification which resulted represented an uneasy compromise between the influence of geology, relief and processes, which was as much implied as understood. Modern geomorphologists have tended to group together all these descriptions of networks and call them morphological classifications, which though interesting, are not sufficiently precise or explanatory to be of use to the modern student of drainage patterns. Some examples of these morphological classifications can be seen in figure 5.11.

Modern approaches to drainage patterns or *networks* stem, as do so many aspects of changing ideas on hydrology, from the work of Horton in the 1930s and 1940s, and they use mathematical methods to describe and analyse drainage networks in order to try to develop laws for the manner in which they function. Some of these mathematical attempts to quantify networks are seen below, but to understand them some basic terminology must be mastered.

The *drainage basin* is the area of catchment of the river and its boundary is marked by the *watershed boundary*. The latter can be difficult to identify in areas of indistinct relief but is generally taken to be a line joining the highest points between different drainage basins, or, failing that in level interfluve areas, the point equidistant between stream sources. Within the drainage basin streams exist in different *orders*. There is some confusion over the numbering of different stream orders but the system most commonly in use is the Horton-Strahler method shown in figure 5.12. Each order stream has its own drainage basin, catchment area, and so on, and it has been found possible to identify a number of laws which seem to operate within the drainage basins of different orders and these are shown below.

Drainage density is a simple expression of the length of stream channels per

Fig 5.9   *The River Tywi meandering through Carmarthen to the sea*

Fig 5.10   *Braiding*

unit area in the drainage basin, and is derived simply by dividing the total length of all stream channels (ΣL) by the area of the drainage basins (A) thus

Drainage density $D = \dfrac{\Sigma L}{A}$

D becomes of major hydrological significance because it reflects land use, and affects infiltration, the basin response time between precipitation and discharge, as well as being of geomorpho-

logical interest particularly for the development of slopes.

The *constant of channel maintenance* is the inverse of drainage density and is a measure of the area needed to support a given length of stream channel and so will indicate the influence of geology, for example, on the drainage pattern.

Constant of channel maintenance
$C = \dfrac{1}{D}$ where D = drainage density

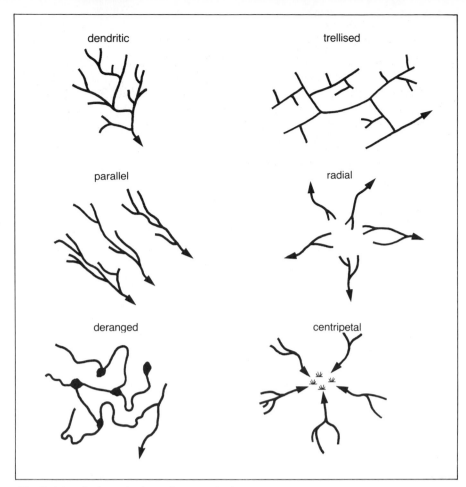

Fig 5.11   *A description of some drainage patterns*

A more sophisticated measure of drainage patterns is the bifurcation ratio, a measure of the amount of branching of a stream network. It is derived by dividing the number of streams of one order by the number of streams of the next highest order. The values gained are then usually expressed as a mean bifurcation ratio which indicates the complexity of the drainage network being examined. A value for mean bifurcation ratio for figure 5.12(a) is derived as follows.

Stream order	Number of streams	r_b
1st	16	
		3.2
2nd	5	
		2.5
3rd	2	
		2
4th	1	
		$\Sigma r_b = 7.7$

Mean bifurcation ratio $\overline{r_b} =$

$$\frac{\Sigma r_b}{3} = 2.56$$

Values for $\overline{r_b}$ are between 3.0 and 5.0 in areas which are not unduly affected by structural control.

Similar to the mean bifurcation ratio is the stream length ratio derived by dividing the total stream lengths of one order by the total stream lengths of the next highest order. This ratio is also worked out for the simple drainage network in figure 5.12(a).

Stream order	Total length (km)	l_r
1st	14.25	
		1.9
2nd	7.5	
		3.3
3rd	2.3	
		2.3
4th	1.0	
		$\Sigma l_r = 7.5$

Mean stream length ratio $\overline{l_r} = \dfrac{\Sigma l_r}{3} = 2.5$

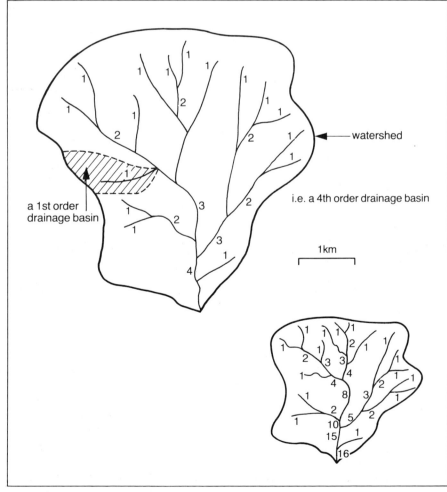

Fig 5.12   *Stream ordering   (a) the Horton-Strahler system*
(b) the Shreve system

The opportunities which are afforded by gathering this sort of data are endless at first glance. Do drainage networks originating on the same geological strata have similar drainage density? Is there a significant relationship between

35

mean bifurcation ratio and a variety of climatic data? Questions such as these were, indeed, intended by Horton to be answered by quantifying stream networks in the manner shown above, then relating them to different environments. In practice there has been some fruitful work done on the topic but the majority of attention has been concentrated on individual stream networks. Perhaps the lack of success in network/environment correlation has been due to two things. First, without strong geological control on drainage networks the number of factors that can influence the eventual network characteristics is so great that the network pattern becomes a randon one from one area to another. Second, it can be argued that the end product of the quantification process represents so much generalising and taking of mean values that the results at the scale at which they are obtained, are too coarse to help understand the detail which really influences network form.

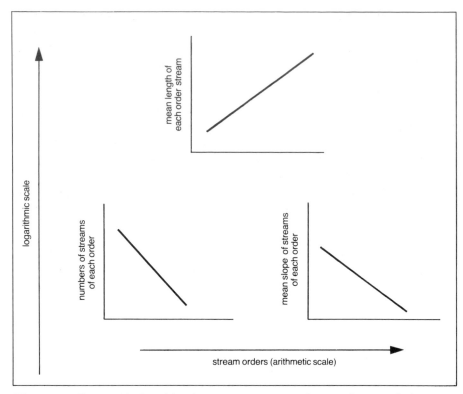

Fig 5.13   Some relationships between stream orders and some drainage basin parameters

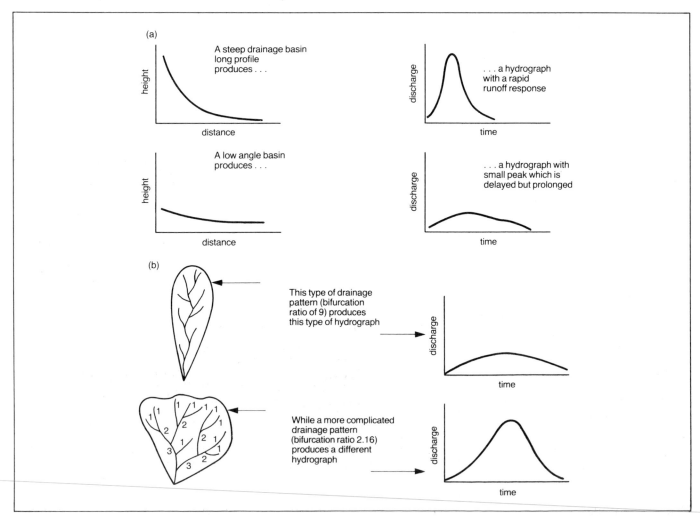

Fig 5.14   Drainage basin characteristics and their effect on hydrograph form

It is not surprising, therefore, that Horton's laws are seen to be most applicable in freely-developing networks (i.e. no overriding structural control) and, even more significantly, stream networks generated by random computing fit Horton's Laws very closely. Figure 5.13 shows the sort of relationships that have been observed to exist between stream orders and various drainage basin characteristics. If there is a criticism of this approach to network analysis it is that perhaps too much effort has been put into applying network data to Horton's Laws and not enough research has been done into *why* drainage patterns are organised as they are in nature.

Basin morphology

It is not only the linear aspects of drainage systems that have been analysed quantitatively. The late 1950s and the 1960s in particular, saw a large number of attempts to define mathematically the character of the entire drainage basin. Shape and relief were two aspects of the drainage basin which received close attention.

Drainage basin shape can be measured in a variety of ways, and three of the standard methods are as follows.

Horton's form factor (F), devised in 1932

$$F = \frac{A}{L}$$

A = drainage basin area
L = longest axis of drainage basin

Miller's circularity ratio (C), devised in 1953

$$C = \frac{A_1}{A_2}$$

A_1 = drainage basin area
A_2 = area of circle with the same perimeter as the drainage basin

Thus, $C = \dfrac{4A}{P}$

P = perimeter of drainage basin

Schumm's elongation ratio (E), devised in 1956

$$E = \frac{\text{Diameter of circle with same area as the basin}}{L}$$

Thus, $E = \dfrac{2.\sqrt{A/\pi}}{L}$

It is also possible to measure shape by comparing irregular shapes (such as drainage basins) with the measured parameter of a known regular geometric figure. For example, comparing the longest and shortest area of a drainage basin with the same axes of an ellipse. Another method is to 'fit' the drainage basin as closely as possible to a hinged 'n'-sided polygon (usually an octagon) but, useful though this method can be, it is not quite as readily applicable as the methods described above which need no more than an Ordnance Survey map, pencil, ruler and calculator, and which have been used by the author to give very consistent shape measure-

ments of drainage basins developed on different geology in Britain, as well as to measure the shapes of upland valleys (e.g. 1st order basins on Dartmoor granite) to see to what extent they are the same shape.

Why bother with measurements of drainage basin shapes? In hydrological terms one answer lies in the influence of drainage basin shape on hydrograph forms (see fig 5.14). But one cannot hope for too clear-cut a relationship because shape, as well as other aspects of the drainage basin, can be influenced by a variety of factors, not least rock type.

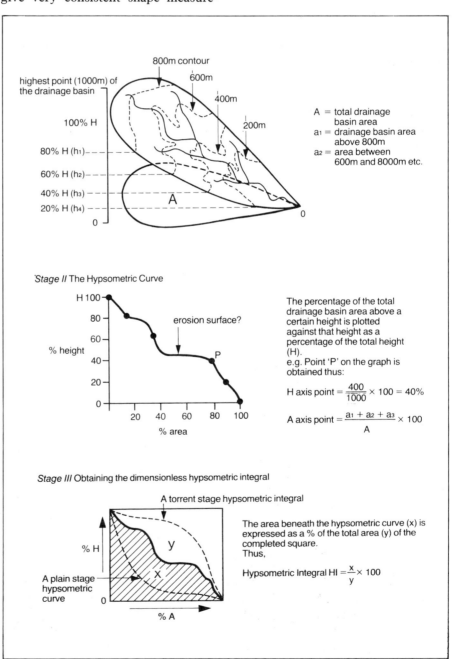

Fig 5.15   The hypsometric integral

Much the same sort of arguments apply to measure relief aspects of the drainage basin. Drainage basin relief will be affected by rock structure in the broadest sense, and the angle of slopes may be partly determined by the stream processes operating at their bases. But, in turn, relief will affect a number of hydrological factors, such as rates of infiltration, runoff, throughflow, the water table surface and so on. Two simple measurements of drainage basin relief are Schumm's relief ratio and Strahler's ruggedness number.

Schumm's relief ratio (1956)

$$R = \frac{H}{L}$$

H = difference between highest and lowest points in the drainage basin

L = horizontal distance along the longest axis of the basin parallel to the main stream

Strahler's ruggedness number (1958)

RG = H Dd

Dd = drainage density

Other relief possibilities include long profiles and cross-profiles of the basin as well as random sampling of slope angles to derive a mean value for basin slope.

In view of the geomorphological as well as hydrological consequences of drainage morphometry it is worth mentioning *Strahler's hypsometric integral*, an expression of drainage basin form which indicates a three-dimensional picture of the drainage basin and which has been used in connection with possible theories of landscape evolution. Figure 5.15 shows how the integral is derived. The penultimate stage, the hypsometric curve, is of special value because it identifies significant areas of flat land within the drainage basin which might indicate past erosion surfaces, while the form of the curve itself can be equated with different stages of landscape evolution. For those who seek to add some objectivity to the Davisian cycle of erosion, the integral has been used to classify landforms: values greater than 66 per cent might be equated with the torrent stage, 33–66 per cent the valley stage and below 33 per cent the plain stage of the fluvial landscape.

EXERCISES

5.3 In the chapter you have just read, some reasons for reduction of discharge are given. Make a list of the features which you think could cause an *increase* in river discharge.

5.4 Study figure 5.16, which is a map of the River Sid drainage basin.
(i) Pebble samples were taken from the River Sid at points 1, 2 and 3. Their mean size and roundness are shown below (the greater the roundness index the more rounded the pebble). What differences can you discern, and how would you explain them?
(ii) (a) Number the streams of the River Sid in the Horton-Strahler manner. What order drainage basin is the River Sid?
(b) Calculate the drainage density of the River Sid basin. Calculate the bifurcation ratio and the constant of channel maintenance for the River Sid.
(iii) Comment on the patterns you observe when, on semi-logarithmic graph paper you plot stream order (x-axis) against (a) the number of streams of each order and (b) the total length of streams of each order.
(iv) Extend the kilometre grid over the entire drainage basin, and mark on the map the watersheds of each order (a start has been made for you). On semi-logarithmic graph paper plot stream order (x-axis) against mean basin area of streams of each order. Comment on the pattern you observe.
(v) Calculate the hypsometric integral for the River Sid. Is the river in torrent, valley or plain stage?

Point	Mean pebble size (mm)	Mean roundness index
1	69	265
2	49	383
3	44	506

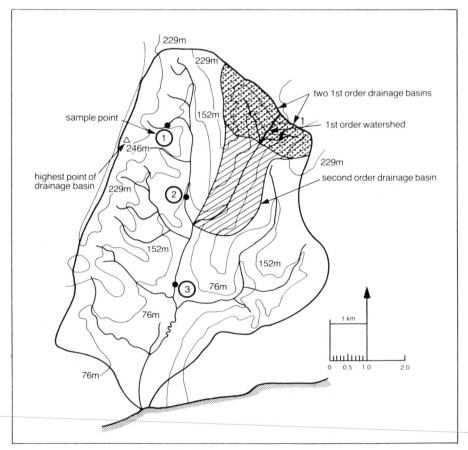

Fig 5.16   The River Sid drainage basin

6 The water authorities of England and Wales

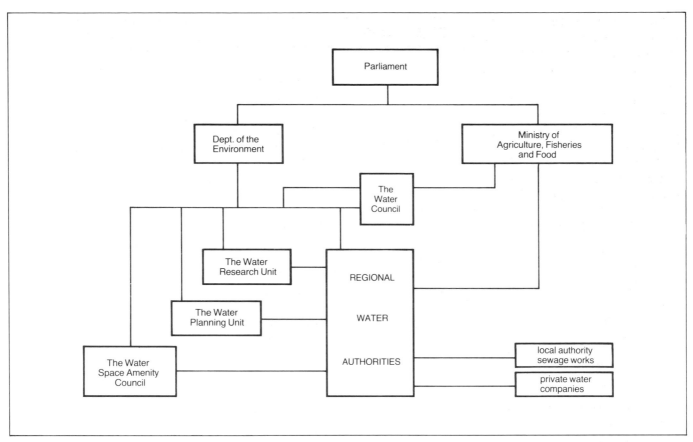

Fig 6.1   The administrative hierarchy for water in England and Wales

Previous chapters have shown how information about water is gained. The next four chapters hope to show how this knowledge is put to use, at a variety of scales, to benefit the human race.

The administrative hierarchy of water in England and Wales is shown in figure 6.1; the main executive work is done by the regional water authorities, but the chief advisory body is the National Water Council which advises government ministries and water authorities, a role similar to that of the more strategically-minded Water Planning Unit.

The Water Planning Unit faces the basic disparities between national demand for water and its availability. The imbalance between the two can be seen in figure 6.2, and it is clear that the areas of greatest demand are removed from the zones of greatest supply. Further, the British climate is notoriously unpredictable, so that *water storage* and *water transfer* are needed to guarantee a public water supply. Fortuitously, effecting this public supply demands a level of control on water which can also be used to predict and prevent floods.

(a) Population

(b) Rainfall

over 200 people/km²   over 1000mm rainfall per annum

20–200 people/km²   600–1000mm rainfall per annum

below 20 people/km²   below 600mm rainfall per annum

Fig 6.2   The distribution of population and rainfall in the United Kingdom

The drought of 1959 prompted the Water Resources Act of 1963. Ten years later, the Water Act of 1973 set up the new regional water authorities shown in figure 6.3, which emphasised the need for a smaller number (10) of administrative units to control and co-ordinate water in England and Wales. The worst drought for 250 years in 1976 spawned the Drought Act, extending the reserve powers of the water authorities.

In chapter 1, we looked at the conflicts of interests with which water authorities have to deal, namely sewage and industrial waste disposal, pollution control, maintenance of fisheries and recreation, flood control and, above all, a plentiful and potable supply of water to domestic and industrial consumers (see fig 6.4).

Water storage

Three possibilities for water storage exist, each having advantages and disadvantages.

(1) Storage in upland areas has the advantage of being in the wettest parts of the country and water is best controlled as soon as it falls. Traditional storage sites, deep narrow valleys, are more readily available, particularly in upland areas that have been glaciated. Upland climates are cool, so losses by evaporation are small, especially if the ratio of surface area to stored volume is large as in deep valleys. Upland geology is generally harder and more impervious than lowland, thus reducing seepage to ground water. Gravity-feed to lowlands is possible, using natural river channels, whose discharge can be better regulated and made less susceptible to flood. Hydro-electric power becomes possible, though pumped storage may be used to offset limited local demand, and there may be a power loss during transfer to the demand. Population densities are low, partly due to the limits imposed by a short growing season, and so disruption to communities is minimal. Finally, drowning upland farmland is a lesser evil than the same deed elsewhere and the ensuing lake will be less prone to pollution and may act as a magnet to tourism.

Pumped storage uses surplus energy, usually at night, to pump water back

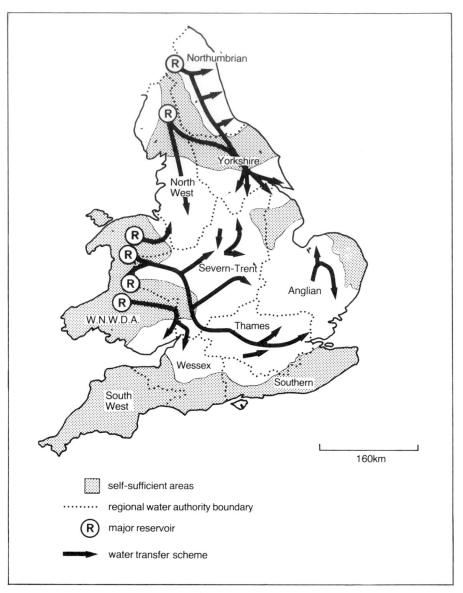

Fig 6.3   Water authority boundaries and major water transfer schemes in England and Wales

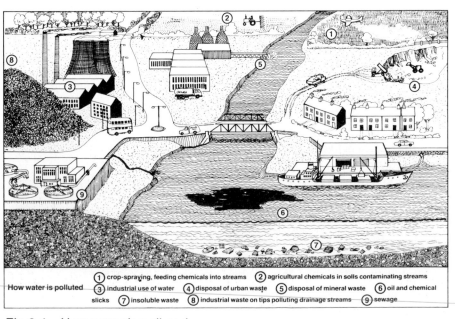

Fig 6.4   How water is polluted

into the storage site so that it can be used again. One of Europe's largest and most spectacular pumped storage schemes is nearing completion at Ffestiniog in north Wales. A more traditional type of upland storage is shown in figure 6.5.

It is possible to object to upland storage. Sites may have costly and difficult access and geological complexity can cause problems for dam construction. Traditional environments can be changed drastically; marshy source areas may be drained so that rapid runoff induces soil erosion; water release downstream can be periodic and sudden, to the detriment of fish stocks such as migrating salmon. Population densities may be low but most people live in the valleys, often in tightly-knit communities, who resent valley land, the best land, being flooded. Moreover, storage sites may be far removed from the demand, requiring expensive aqueducts and pumps.

(2) Storage in lowland areas has the immediate attraction of being close to demand, easier of access and more readily used for multi-purpose developments, including recreation, whose facilities improve public relations (see fig 6.6). Unfortunately, drowned land may be good farming land and significant compensation paid to people displaced. Transfer by gravity means is limited, pumping stations may be needed, and there is negligible opportunity for power generation.

(3) Water can be stored artificially beneath the ground, and recent years have seen a movement towards re-charging aquifers by pumping water into them. Evaporation losses are reduced, discharge in water courses becomes more constant, and ground water sources can be tapped with fewer consequences. The method is attractive to a city such as London whose water table has been lowered dramatically over the years. This type of water storage demands detailed knowledge of sub-surface geology and water movements (see fig 6.7).

Upland storage has been the most common in the United Kingdom, and some dams are having to be enlarged as the best sites become used up. New sites face increasing environmental

Fig 6.5 Upland storage at Pitlochry, Perthshire

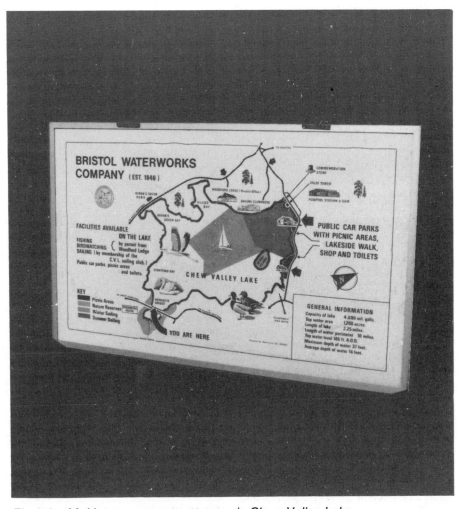

Fig 6.6 Multi-purpose water storage in Chew Valley Lake

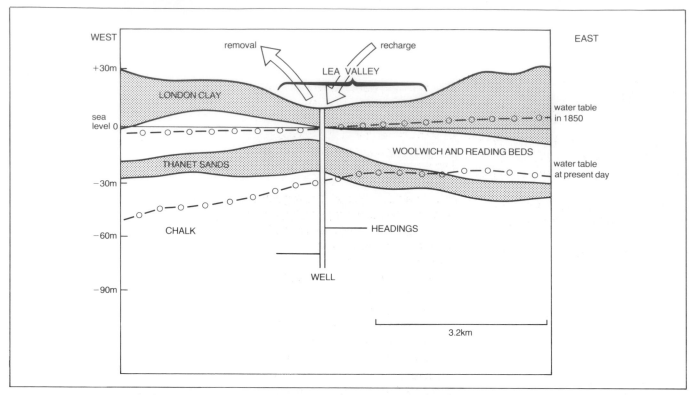

WEST   EAST

+30m

LONDON CLAY

sea
level 0   water table
                                                       in 1850

THANET SANDS   water table
                                                       at present day
−30m

WOOLWICH AND READING BEDS

CHALK   HEADINGS
−60m

                         WELL

−90m

LEA VALLEY   removal   recharge

3.2km

Fig 6.7   London's falling water table – but water can not only be taken from
the well, it can also be pumped back in to recharge the aquifer

opposition; Cow Green in Teesdale, a
site of special scientific interest because
of its Ice Age flora, faced a vigorous
opposition campaign before being
flooded to meet the demands of Tees-
side. New reservoirs therefore tend to
be lowland and large; Chew Valley,
Grafham and Rutland Water are three
typical lowland sites.

Water transfer

Figure 6.3 also shows a simplified
version of the major long-distance
water transfers in England and Wales.
The pattern is predictable. Wet uplands
become storage areas from which water
is led by combinations of natural
channels and aqueducts to areas
deficient in supply. The mid-Wales
dams, for example, store water which
may eventually be consumed by
thermal power stations in the Trent
valley, industrialists in the west
Midlands, or the public of London.
Ground water sources are also tapped
for transfer; for example, the chalk of
the Thames basin or East Anglia, and
ultimately a national water grid may
emulate the national electricity grid.
Such a grid would increase greatly the
flexibility of water supplies without
any net increase in the volume of water
available. Such developments would
aid flood control as well. Floods are
looked at in more detail in chapter 8.

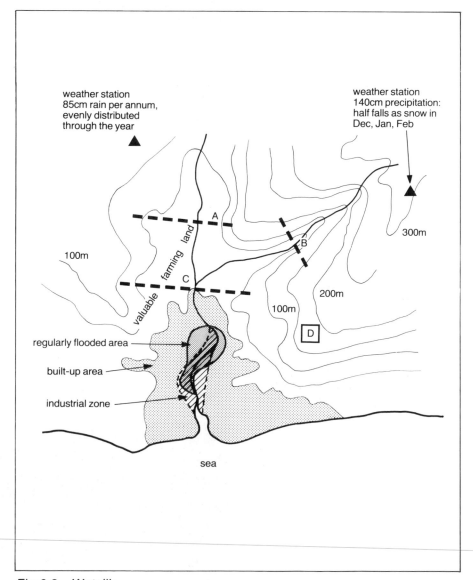

Fig 6.8   Wetville

EXERCISES

6.1 How reliable is the rainfall in your area? Rainfall reliability can be calculated by the following technique. Variability less than 10 per cent means that your weather station experiences fairly reliable rainfall.

6.2 In your home area, identify the major users of water and the major sources of supply. Write a short essay to identify any problems which might exist in reconciling the two in view of seasonal fluctuations in supply or demand.

6.3 Study the map of proposed water transfer in England and Wales (fig 6.3). Find out which reservoirs are involved in the scheme, and which demands the scheme is trying to supply.

6.4 Building dams and creating reservoirs usually causes controversy. Choose a newly-created reservoir in the United Kingdom and assess the arguments both for and against the scheme.

6.5 The city of Wetville is regularly flooded. You are the hydrological troubleshooter brought in to advise the local authority on possible remedies. These are:
(i) to relocate people and industry from the flooded areas to sites on high ground at D;
(ii) to build dams at either A, B, or C. What would your comments be? You should bear in mind the most complete answer, which would take into account the most beneficial and least harmful side effects of each proposal. Would you want to know any other information before making your comments?

Year	Total rainfall (Y) (Look up the rainfall totals for your area)	Mean annual rainfall for the last 35 years (X) (This figure will be constant)	Difference between Y and X
1976		X	
1977		X	
1978		X	
1979		X	
1980		X	
1981		X	
	Mean annual total equals $\Sigma\ \dfrac{Y}{6}$	X	Mean difference from mean annual rainfall

$$\% \text{ variation} = \frac{\text{mean difference}}{\text{mean annual total}} \times 100 = \qquad \%$$

7 The Wessex Water Authority

The Wessex Water Authority (WWA) was inaugurated following the 1973 Water Act. It administers the area shown in figure 7.1 which contains a resident population of approximately 2 350 000 people, increasing at 1.05 per cent per annum since 1966. At peak holiday periods the population of selected zones within the area can increase by up to 500 per cent. Within the authority's area there exists a diversity of water supply sources and demands which present a correspondingly wide variety of problems.

To meet these problems the WWA has a clearly-defined administrative hierarchy shown in figure 7.2. The Authority has a chairman appointed by the Secretary of State, four members appointed by the Avon, Somerset, Dorset and Wiltshire County Councils, four by District Councils, four by the Secretary of State to be responsible for such things as water for industry, two by the Ministry of Agriculture, and two members to represent the urban concentrations of Bristol and Bournemouth/Poole. The Authority is advised by the National Water Council, and its policies are implemented at divisional level by three divisional directors, each heading a team to cover all aspects of water management. The WWA personnel total 2 280 people, and the Somerset division has 624 of these. The main concerns of the Authority are water supply, sewage disposal, catchment management, land drainage and flood control, and providing recreational facilities. Guidelines are laid down by the 1973 Act, but increasingly the Authority is moving towards directives from the EEC.

There are six broad EEC directives. They are concerned with the quality of surface water for drinking; the discharge of dangerous substances to the aquatic environment; quality of bathing water; quality of fresh water to support fish life; quality of water for human consumption; and the quality of water for agricultural use, shellfish growth, and the protection of ground water quality. Full implementation of the EEC standards is likely to have considerable financial consequences for the Authority (see fig 7.3).

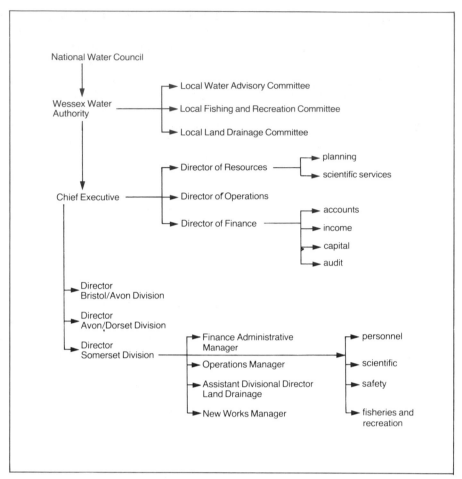

Fig 7.1   The Wessex Water Authority

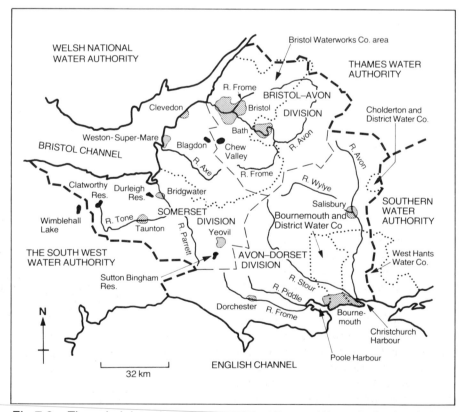

Fig 7.2   The administrative structure of the Wessex Water Authority: the structure of the Bristol/Avon and Avon/Dorset divisions are similar to that of the Somerset division

Water supply

Major (national) legislation

The Reservoirs Acts of 1930 and 1935
The Water Acts of 1945 and 1973
The Public Utilities Street Works Act, 1950
The Water Act, 1973
The Drought Act, 1976
The Water Charges Equalisation Act, 1977

Wessex water supplies are provided by the combined efforts of the Authority and private water companies. Both cater for an anticipated 2.5 per cent per annum increase in consumption, and the relative shares of total peak day consumption are shown below. They supply 96 per cent of Wessex population connected to the public water mains.

Table 7.1 Peak day consumption within the Wessex Water Authority

Management unit	Peak day consumption ml/day
Wessex Authority	364
Bournemouth and District Water Co.	73
Bristol Waterworks Co.	313
Cholderton and District Water Co.	1
West Hants Water Co.	46

One ml/day = 0.22 million gals. per day

To form the type of mains network shown in figure 7.4, several formerly independent systems have been merged to create a more reliable service, but the Authority accepts that some isolated districts will never be part of a regional grid. The 1976 drought encouraged development of several hitherto unexploited water sources which have since become permanent features to meet peak demands or future drought conditions. They were used in the autumn of 1978 in Somerset when only 27 per cent of normal rain fell, giving concern that the Authority's reservoirs would be insufficiently replenished to meet usual demands. Of the potable water consumption 50 per cent comes from surface sources, 50 per cent from sub-surface. All sources are chlorinated, but surface waters receive extra treatment and, significantly, the Authority's five major water treatment plants are in the Somerset division which has the highest proportion of poor-quality surface water. The Somerset division completed in 1979 a project to transfer good-quality water from Wimbleball reservoir in the South-West Water Authority, via the mains grid system, towards the demands of Taunton and Yeovil. Other Authority schemes intend to increase water supply to the CEGB power station at Hinkley Point, and to the popular tourist resorts of West Somerset and South Dorset.

The Wessex Plan for Water 1979–84 has four aims for water supply – to link existing systems to increase flexibility of water use, to install new reservoirs and pumping stations where appropriate, to strengthen the distribution system, to improve efficiency by reducing losses from the existing supply system, and work towards the standards laid down by the EEC.

THE EUROPEAN WATER CHARTER

1 There is no life without water. It is a treasure indispensable to all human activity.

2 Fresh water resources are not inexhaustible. It is essential to conserve, control, and wherever possible, to increase them.

3 To pollute water is to harm man and other living creatures which are dependent on water.

4 The quality of water must be maintained at levels suitable for the use to be made of it and, in particular, must meet appropriate public health standards.

5 When water is returned to a common source it must not impair further uses, both public and private, to which the common source will be put.

6 The maintenance of an adequate vegetation cover, preferably forest land, is imperative for the conservation of water resources.

7 Water resources must be assessed.

8 The wise husbandry of water resources must be planned by the appropriate authorities.

9 Conservation of water calls for intensified scientific research, training of specialists and public information services.

10 Water is a common heritage, the values of which must be recognised by all. Everyone as the duty to use water carefully and economically.

11 The management of water resources should be based on their natural basins rather than on political and administrative boundaries.

12 Water knows no frontiers; as a common resource it demands international co-operation.

Fig 7.3   The European Water Charter

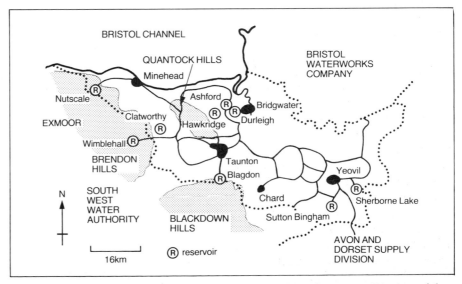

Fig 7.4   The principal public supply network of the Somerset Division of the Wessex Water Authority

Sewage disposal

Major legislation

The Public Health Act, 1936 and 1961
The Public Health (Drainage of Trade Premises) Act, 1937
The Water Act, 1973

Wessex has some 8850km of sewers in varying condition, many serving as combined foul and surface sewers which also contain storm water overflows which discharge to local water courses. The dates of sewer construction are shown in Table 7.2 and there is considerable concern that the different systems are inadequate to cope with the demands made on them. This is particularly so in the case of the combined sewer storm overflows, whose discharges to water courses can create physical, chemical, bacteriological and aesthetic problems that prove unpalatable to the public. Of the 805 storm overflows in Wessex 578 are reckoned to be 'unsatisfactory' or 'questionable'. The Authority's sewer system serves 2 100 000 people, 160 000 by septic tanks or cesspits. The Authority disposes of 610ml per day of sewage, half of which is treated before discharge to rivers; 10 per cent is discharged partially treated to non-tidal rivers; and 40 per cent to coastal waters including the semi-enclosed and picturesque Christchurch and Poole harbours. Sewers may be infiltrated by soil moisture or ground water. Figure 7.5 shows such infiltrations in the Salisbury area, and they increase significantly the cost of treatment plant by taking up sewage capacity.

The Authority is acutely aware of the sewage problem and is currently studying how to improve its service. Its 1979–84 plan aims to protect water courses from stream overflows, to convey safely dirty domestic and industrial water to a suitable disposal point, to convey clean surface water to an adequate water course, to install separate foul and surface sewers on all new urban development and work towards separate systems where dual ones currently exist, to treat sewage to a standard compatible with its likely future use, and to treat sewage in a smaller number of larger plants where economy, environmental protection and the fertiliser and energy potential of sewage can be best realised.

Table 7.2

Date of construction	Foul or combined sewers (km)	Surface water sewers (km)	Total
Pre- 1900	540	120	660
1900–1915	950	240	1190
1915–1945	1860	740	2600
1945–1970	2450	730	3180
Post 1970	910	310	1270
Total	6710	2140	8850

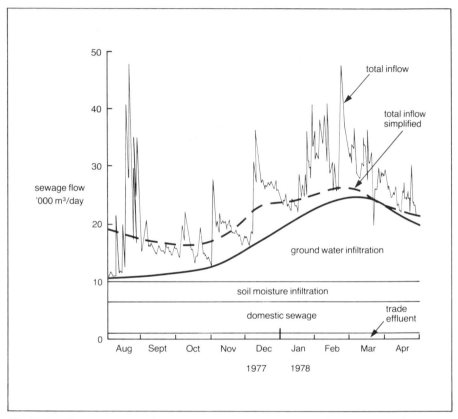

Fig 7.5 Components of total sewage inflow at the Salisbury sewage works during the winter of 1977–8

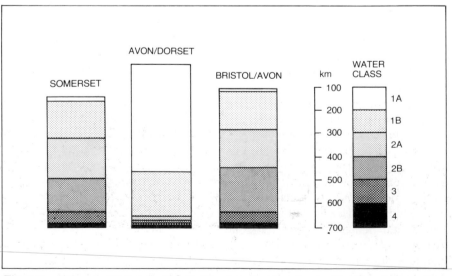

Fig 7.6 Wessex Water Authority: the length of watercourses in each water quality class

Catchment management

Major legislation
The Rivers Acts, 1951 and 1961
The Clean Rivers Act, 1960
The Water Resources Act, 1963
The Water Act, 1973
The Control of Pollution Act, 1974

The 1963 Water Resources Act introduced a licensing system before water could be abstracted from or added to surface and sub-surface waters. The WWA is licensed to abstract 1072ml/day for public supply, to import a further 134 ml/day from neighbouring authorities, and export 43 ml/day to adjacent regions. Licensed abstraction quotas are seldom needed as Table 7.3 shows.

Abstractions from and additions to surface waters have an effect on Wessex rivers. Using criteria laid down by the National Water Council, the WWA analyses river-water quality and sets long and short term objectives for its standards. Within Wessex, Avon and Dorset have by far the greatest length of top-grade water courses, while Somerset and Bristol/Avon are relatively poorly off. Figure 7.6 shows the length of water courses in each subdivision in each water quality category. Criteria for two sample criteria are as follows:

Grade 1A: dissolved oxygen saturation (DOS) greater than 80 per cent; biochemical oxygen demand (BOD) less than 3mg/litre; ammonia content below 0.4mg/litre; non-toxic to fish; the water is suitable for all abstraction purposes, can support high-quality game fisheries, and is of high amenity value.

Grade 4: watercourses have DOS below 10 per cent; BOD greater than 17mg/litre; the water is likely to be anaerobic; fish are absent or sporadically present; the chief use of the water could only be for low-grade industrial abstraction.

Table 7.3 Licensed abstractions for 1977 (ml/day) (actual abstraction in parenthises)

	Public supply	Industry	Agriculture
Surface water	565 (363)	377 (272)	91 (89)
Ground water	507 (310)	53 (26)	219 (219)
Total	1072 (673)	430 (298)	310 (308)

Land drainage and flood control

Major legislation: The Land Drainage Act, 1976

The WWA is involved with all the obligations of the 1976 Act, namely to improve the channels and limit flooding in the main rivers; to supervise minor streams and land drainage for agriculture; to operate flood prediction and protection schemes including sea defences where necessary. To achieve this, the Authority's planned capital expenditure 1979–84 averages £3.058 million per annum. Typical schemes are flood alleviation in the urban catchment of the River Frome in Bristol; or reconstructing the sea defences at Clevedon and Kingston Seymour; or longer-term schemes such as reducing flooding in the River Parrett lowlands of Somerset to possibly include a tidal barrage.

Already in operation are sophisticated flood warning schemes. Information on rainfall, soil moisture, recorded discharges for strategic points in flood-prone catchments is fed into computers. By and large the flood warnings are useful, but despite close liaison with the Meteorological Office, the 10cm of rainfall that produced the disastrous flooding during May of 1979 in Somerset proved unpredictable. Unfortunately, even if the precise time, duration and amount of rainfall could have been foreseen, few effective precautions could have been taken; antecedent rainfall had produced soil moisture near field capacity, and watercourses so full they had little storage capacity to spare to absorb the large-scale runoff.

Recreation and amenity

Major legislation: Salmon and Freshwater Fisheries Act, 1975

Fishing is very important within Wessex. Blagdon and Chew Valley reservoirs, maintained by Bristol Waterworks Co., are justly famous for their trout fishing, and both are regularly re-stocked. Equally famous are the wild trout waters of the River Avon, Frome, Wylye and Piddle chalk streams, and very different high-quality coarse fisheries are the Dorset Stour, or the myriad of water courses in the Somerset levels. In fisheries maintained by the Authority charges are levied to cover the costs of maintaining fish stocks and policing the waters. Coarse fish for re-stocking are often reared in lagoons at the major sewage works; the fish grow phenomenally quickly and are useful to offset losses to pollution. In many cases land adjacent to reservoirs provides picnic areas, nature trails, bird-watching and is leased to sailing clubs.

The Authority also maintains navigable waterways, the River Avon between Bath and Bristol, the Kennet and Avon and Taunton and Bridgwater canals, and, supporting industrial archaeology, the Authority is restoring part of the derelict Westport Canal.

The Wessex Water Authority is clearly involved in a wide variety of activities. Figure 7.7 indicates the allocation of investments for 1979–80 within the Authority.

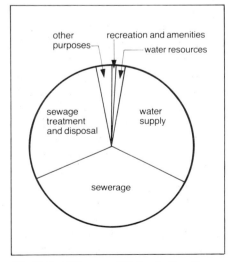

Fig 7.7 Wessex Water Authority: allocation of capital investment for 1979-80

7.1 Complete the linkages in figure 7.8 which shows a simple system of conflicting and complementary water uses. Use broken lines for complementary uses and unbroken lines for conflicting uses. A start has been made for you.

7.2 How much do you pay in water rates? What proportion of the total rates you pay does it represent? Do you think you are getting value for money? Would you prefer to pay for water as you use it (metered water) or in relation to the rateable value of your house?

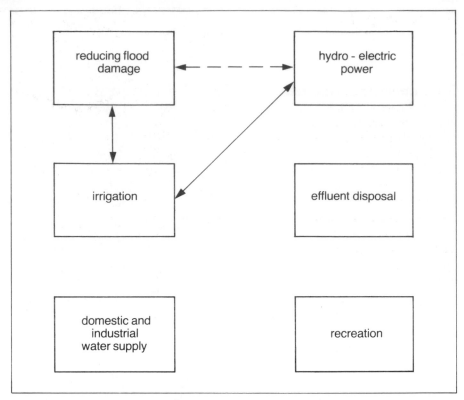

Fig 7.8   Conflicting and complementary uses of water

8 Floods

Floods are ideal phenomena for study by the modern applied hydrologist. Their causes are intimately bound up with types of precipitation, infiltration, runoff in a variety of forms, channel characteristics and the man-affected pattern of land use. Their consequences can also be physical in nature but more dramatically can take life and cause large-scale social and economic hardship; uncontrolled flooding can represent a waste of a much-needed resource.

Many of the serious problems associated with floods and flooding are attributable to the fact that man has always been attracted to low-lying lands, occupying them either through ignorance of the likelihood of floods and their consequences, or because he is prepared to take a calculated risk that his protective measures will save him from all but the most catastrophic events. Nowhere is this more obvious that in the occupation of the Hwang Ho flood plain in China, a river, the failure of whose levées extending high above the surrounding countryside, make disaster involving thousands of lives almost commonplace. The same attitude is true of those occupying the great delta comprising most of Bangladesh. Pressure of population, scarcity of land and the rich sediment deposited by the Ganges make the delta attractive to settlers, yet recent decades have seen a succession of catastrophic floods taking a very high toll of lives, the most recent being in 1974 when approximately 6 million people were left homeless.

Even advanced societies do not escape. 1953 saw much of the east coast of England and the Low Countries flooded; freak tides caused flooding in Queensland in 1974 when between 600 000 and 700 000km² were flooded; and in the United States regular flood warnings may save lives but cannot always prevent damage to property in the wake of cyclones moving northwards from their breeding ground in the Caribbean. Mankind can even be accused of creating flood risk by constructing dams that subsequently fail (fig 8.1). An undetected geological weakness caused the Malpasset Dam in southern France to crumble, releasing a fast-moving wall of water that

Fig 8.1   The power of floods: the aftermath of the Morvi dam failure (India)

destroyed most of the town of Frejus in 1959.

From the above it can be seen that the causes of floods are numerous, but four main categories can be recognised.
(1) heavy rainfall
(2) rapid snow and ice melt
(3) coastal floods
(4) geophysical catastrophic events, including dam failure
In each case the effects of flooding meet the types of response by man shown in figure 8.12.

Floods due to heavy rainfall

On a world scale heavy rainfall is the most important cause of large-scale flooding; pages 51 to 53 show the relationship between the nature of rainfall and runoff. The degree of flooding following a given amount of rainfall will depend on antecedent moisture conditions in the drainage basin, that is the extent to which geology is saturated and soils at field capacity, the morphometry of the basin (its size, shape and drainage network

characteristics), and the permeability of the drainage basin surface including geology and land use. These will all affect the form of the flood hydrograph (see fig 4.11).

It is difficult in the short term to predict rainfall characteristics and some floods are akin to the flash floods of arid areas, but fortunately such floods are limited in extent, rarely blanketing entire drainage basins, though they are capable of causing severe damage to first-order drainage basins in hilly areas. Over a slightly longer time-period of several weeks multiple floods can be recognised, perhaps the result of a series of depressions passing over a drainage basin continually overtopping the storage capacity of the drainage basin. On a still longer time scale, seasonal floods can be identified which may last for several months. The Nile is the classic case of such a flood, but similar examples can be seen in all the monsoon and savanna lands of the world. Indeed, the floods already mentioned which ravage Bangladesh are associated with the monsoon-cyclone system, with the remainder of the year having winter monsoons, bringing problems of a different sort – drought!

In British terms there is a large number of well-documented floods. For example 130mm of rain fell in the night of 10–11 July 1968 in the Mendip area and caused considerable local havoc in nearby Pensford and Keynsham as well as in the more immediate Mendip hills (see fig 8.2). Fortunately Chew Valley reservoir was able to retain 1820 million litres of water during the main period of runoff (the lake rose 42cm in less than 12 hours) and prevented much serious damage in the Chew river basin (see fig 8.3).

The West Country seems to be one of the most flood-prone areas of Britain, and major floods have occurred at regular intervals, but the most famous remains the Lynmouth flood of 1952 (see fig 8.4).

The East and West Lyn rivers drain small catchments on the north slopes of Exmoor before entering the sea close to each other in Lynmouth Bay. The geology of the area, particularly to the west, is Devonian sandstone and slate, not very permeable and supporting only thin soils. The Lyn valleys are steep, narrow, wooded and picturesque. On the night of 15–16 August 1952, up to 300mm of rainfall fell on the southern watersheds of the Lyn basins, following a period of two weeks when it rained twelve days out of fourteen, which ensured the catchments were already nearing storage capacity. This additional rainfall, thought to be from a deep depression from the Atlantic, ran off the surface of Exmoor with great rapidity into the East and West Lyn rivers, so that their discharge exceeded the greatest recorded discharge for the Thames at Teddington Lock. Rapid and extensive erosion took place and the eroded debris, boulders and trees accumulated to form temporary dams in the narrow valleys whose collapse contributed to the flood waves that devastated the tourist resort of Lynmouth, killing 34 men, women and children and causing estimated damage of £9 million. Two salutary points are first that some fluvial deposits in the catchments were untouched, suggesting a greater flood

Fig 8.2   Flooding at Cheddar Gorge, Mendip Hills, July 1968

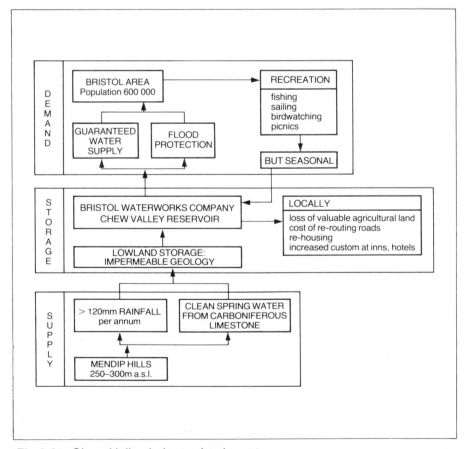

Fig 8.3   Chew Valley Lake: a simple system

in the past, and second it is calculated that an event of that magnitude is unlikely to recur for 50 000 years!

Floods due to snow and ice melt

Floods due to snow or ice melt are most likely to occur in high altitudes, particularly in high mountains which accumulate snow. The important factors are the thickness of snow, the rate at which it melts and whether the ground beneath is frozen or not. Gradual melting, even of great thicknesses of snow, will only lead to a sustained high river discharge. Sudden melting may produce large floods whose effects may be exacerbated if ice floes form temporary dams which suddenly give way.

Causes of rapid melting are primarily sudden climatic change such as quick transition from winter to spring with its attendant high temperatures. Snow melt can be exceptionally fast if caused by rainfall which also makes an additional contribution to the runoff. The great north-flowing Siberian rivers cause vast annual flooding because their downstream (northerly) courses remain frozen while melting has occurred further upstream (to the south). The rivers flood extensive areas of the taiga and tundra in this way. Other factors may also influence rapid snow melt; for example the aspect of the drainage basin (more rapid melting in those that face south) or the type and distribution of vegetation cover.

Within Great Britain snow-melt floods do occur, particularly in Scotland, but can also occur elsewhere, and some of the most recent were the high flood levels resulting from the unusual heavy snowfalls in Devon, Dorset and Somerset during the Spring of 1978. Ice-melt floods in Great Britain are very rare; we seldom have any large-scale ice, but even in areas which do experience large amounts of ice such as glacial areas in Europe, the rate of ice-melt is rarely rapid enough to cause regular floods, and though occasional catastrophies do occur, a prolonged higher discharge is the more likely result.

Coastal floods

Coastal floods are usually caused either by unusual high tides or by storm surges that produce abnormally high water levels. High tides are most destructive when they become increasingly

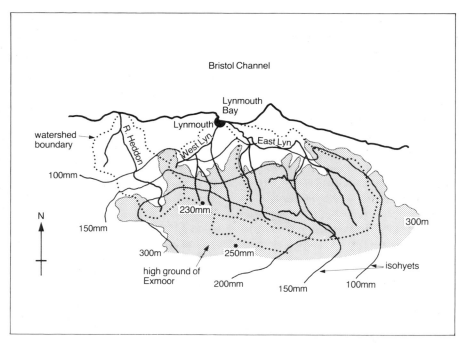

Fig 8.4   The catchment area of the Lyn rivers

Fig 8.5   The Severn bore

confined into narrowing estuaries, and when they hold back river flood waters which then inundate adjacent land. A classic example of the former is in the Severn estuary where tides become increasingly confined, raising the water level abnormally; at the Milford Haven end of the Bristol Channel the tidal range is approximately 4 metres, but at Avonmouth over 14 metres. Sometimes the increase in water height in spring is sufficiently rapid to produce the Severn 'bore', a fast, up-stream moving wall of water (see fig 8.5).

The Thames illustrates the effect of high tides holding back river flood waters and the occurrence has become sufficiently common, and the consequences of flooding London so potentially catastrophic that the Thames Barrage at Greenwich is an attempt to reduce the risk. Similar thoughts lie behind the Dutch Delta Project on the other side of the North Sea.

The North Sea is famous for its storm surges, phenomena defined as a difference between the predicted tide levels and those experienced, and they can be very damaging because so much adjacent land is low-lying. North Sea surges are usually due to very strong winds from the north-west, and the famous surge of 1953 (others have occurred since in 1969 and 1976) was due to the powerful north-westerly winds which were associated with a deep depression moving NW–SE across the North Sea. Meteorological maps for 31 January 1953 show average wind speeds of 40 knots, sufficient to drive water southwards faster than it could escape via the Straits of Dover (see fig 8.6). Fortunately river heights were normal during this period or even more damage might have been done, but even so tidal heights of 2 metres above normal occurred along the English coast from Lincolnshire southwards, while tides over 3 metres above normal occurred along the Dutch coast. Nearly 1900 lives were lost because of the consequent flooding in the two countries, and over 2500 km² of land were flooded, some of it with long-term consequences, by saline water.

Some sea surges occur on open coastlines, but these are usually because of high tides or waves generated by very strong winds in the open oceans. Examples include the storm waves, generated by Caribbean hurricanes, advancing on the southern and eastern coasts of the United States, such as hurricanes David and Frederick in September 1979 (figures 8.7 and 8.8).

Geophysical catastrophies and dam failures

Some of the most damaging of all floods occur in coastal areas which are either close to, or exposed to the effect of, disturbances in the earth's crust. Volcanic eruptions, earthquakes or landslides caused by the two effects can generate vast shock waves that can travel thousands of miles through the oceans. Figure 8.9 shows the generation of a tsunami. Tsunami are a sequence of fast-moving waves generated by ocean floor disturbances such as earthquakes. In the open ocean, moving from the epicentre of the disturbance, they cause little damage and may even be unnoticeable, but when they reach shallow coastal waters their height increases rapidly with relatively little reduction in velocity, and the damage caused can be enormous. Seismographs can give a little warning of approaching tsunami, but more important is a knowledge of offshore

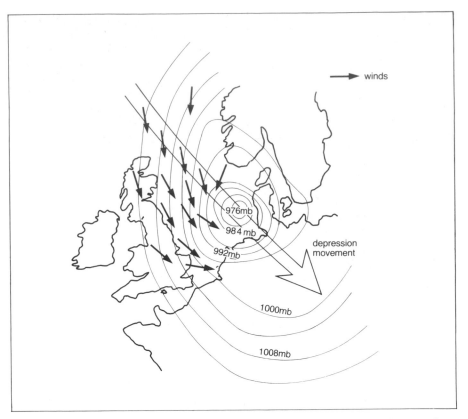

Fig 8.6   Ideal conditions for a North Sea surge: a deep depression moving NW-SE through the North Sea producing strong north-westerly winds

Fig 8.7   Hurricane formation off the coast of the USA . . .

sea-bed configuration, for it is this which largely determines the height of the waves which advance towards land. The worst tsunami on record is that of 1876, for which no amount of warning could have helped, which killed nearly 250 000 people in the lowlands of coastal Bengal. The famous volcanic explosion of Krakatoa off Sumatra in 1883 caused great damage locally, and

its shock waves, transmitted through the oceans, reached Great Britain in only 32 hours, suggesting a tsunami speed of some 650 km/hr.

Fortunately tsunami are relatively rare and so, fortunately, are failures in dams or other man-made water-control structures. When these do occur, they are normally because of unusually heavy precipitation, landslides into the

Fig 8.8   . . . and the resulting storm damage

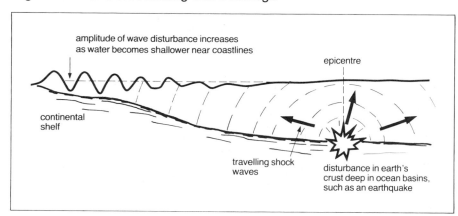

Fig 8.9   The generation of tsunami by tectonic disturbance

retained lake, or by seepage undermining the dam structures; the results are very serious because of the sudden release of vast volumes of water, without warning, on to the people, farmland and industry gathered downstream of the dam. The consequences of dam failure raise the issue of whether they should be even contemplated in areas of tectonic instability. The failure of the Malpasset Dam above Frejus might be one such example and it is possible that a geological disturbance may have urged the landslide into the Vaiont Dam in Italy in 1963, which, despite the dam remaining intact, led to water flowing over the dam and killing over 3000 people downstream. Fortunately each dam failure sheds more light on why dams in general fail, and so knowledge of correct dam structure increases to make failures less likely in the future.

Above we have seen the main causes of floods. Hydrologists may be interested in their geographic effects, for one flood may bring about more change in a landscape than years of 'normal' flow, but it is with predicting and preventing floods that the hydrologist will be most concerned.

Flood prediction

The socio-economic cost of floods to human life, farmland, communications and industry makes it clear that some form of flood prediction is essential. Prediction of the tsunami/dam failure floods are much more difficult than the surge floods, which in turn are harder to predict than the river floods, and it is in the latter field that great advances have been made in flood prediction. There are several approaches which the hydrologist can take to flood prediction.

(1) Some predictions try to predict flood magnitude by *relating it to measured characteristics* of the input into the drainage basin (precipitation) and characteristics of the drainage basin to which values have been apportioned. These are called *deterministic predictions*. The problems which exist lie first in identifying the elements most responsible for flood magnitude, second in deriving a formula into which they can be fitted, and third in making sure that a system exists to enable values for the data to be collected early enough to give a flood warning. Despite its name this method is less than truly deterministic because it does not, for obvious reasons of complexity, take into account all flood-affecting factors, only the most important ones and the ones which can be measured. The US Department of Agriculture uses the following formula for flood prediction.

$$Cl\,max = \frac{C\,A\,Re}{Tp}$$

Cl max = Max. flood discharge (cusecs)
C   = A coefficient, usually 484
A   = Area of drainage basin
Re   = Rainfall excess
Tp   = Time of rise to the drainage basin hydrograph

In Great Britain, J. Rodda, has proposed the formula

$$Cl = 1.08A^{0.77}\,R^{2.92}\,D^{0.81}$$

Cl = mean annual flood
A = Area of drainage basin
R = Mean annual daily rainfall maximum
D = Drainage density

The majority of such formulae can rarely be used freely in all circumstances, being most suitable for small drainage basins with uniform surfaces, such as urban areas, and their coefficients have to be changed from one situation to another. One single formula for all situations would have to contain such a strong stochastic (random) element that it would become vitually useless.

(2) The second major type of flood prediction is the method based on *statistical probability*, namely that there is a *recurrence interval* for floods of a certain magnitude. In other words, the larger the flood, the longer the time

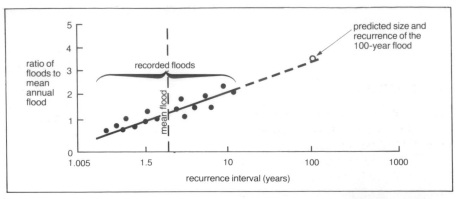

Fig 8.10   Plotting flood magnitude and frequency by the Gumbel method

interval between another flood of similar size. Unfortunately drainage basins do not adhere to probability theory and today's flood with a recurrence interval of fifty years may recur next week, but the *probability* that it will do so in a week's time is less than that it will recur in five years; there is a greater probability for fifty years, and a very strong probability that the fifty-year flood will occur in 500 years. The most famous name associated with this form of flood prediction is that of E. Gumbel, who, on specially-designed Gumbel graph paper, plotted the recurrence interval of extremes of discharge (see fig 8.10). A weakness of the probability method has already been indicated. Others stem from the lack of accurate historical records for the size of past floods, many of which have to be interpolated from geomorphological evidence. In addition, the statistical method assumes all flood causes (rainfall, snow-melt) to have the same effect. The time scale involved for getting a wide range of flood discharges is such that there is no guarantee that climates have not changed in that period. Perhaps the greatest value of probability prediction is in the design of flood prevention structures where the anticipated life of, say, a dam can be weighed against the cost of its construction and the floods it may have to withstand.

(3) Akin to the Gumbel-type probability prediction are predictions of *maximum possible precipitation* (PMP) and *maximum possible floods* (PMF) capable of occurring within a catchment area. Were such predictions possible they would be of great value to the hydrologist and flood control, but each is hard to define with numerical precision and it is almost as hard to relate the two. When correlations are attempted it is usually by using the unit hydrograph method mentioned above. Successfully relating PMP and PMF would be enormously valuable to those trying to prevent flooding, and define flood-risk zones.

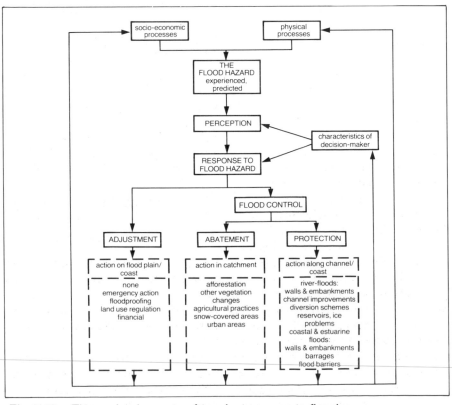

Fig 8.11   Coping with flooding

Fig 8.12   The main elements of man's response to floods

Flood damage

Despite all precautions floods still occur. It is hard to assess flood costs in terms of ill-health, low spirits or anxiety, but more accountable losses are those of production in industry and agriculture, and the direct losses such as damage to houses, cars, furnishings and so on (see fig 8.11). The financial cost of flooding is greatest in affluent societies, but these have the benefits of well-organised assistance from police, firemen, the armed forces, local government workers and not least the insurance companies. Monetary consequences may be less in underdeveloped countries but the human cost is likely to be much greater. Warnings were given of the Bangladesh floods mentioned earlier and of the Morvi dam failure in India in August 1979; most people preferred to stay where they were, perhaps because of lack of transport and nowhere to go, but more likely because of anticipated benefits in terms of land and international aid when the floods were over. Perception of flood hazard is therefore important; unless people realise they are at risk they are unlikely to take action to reduce flood consequences and perhaps more research should be made into the most effective methods of communicating flood warnings (fig 8.12).

Flood prevention

Man can both increase and lessen the risk of flooding. Increased risk is brought about by the changes he brings to land use. The deforestation of catchments, especially on steep slopes, can cause accelerated runoff. So, paradoxically, can afforestation in the period before trees have matured, when the drainage patterns lead to a faster runoff. Converting land to arable use can have the same effect, though sensible strip cropping and contour ploughing can reduce the flood peak; similar ideas of block planting conifers in chequerboard fashion is being tested in North America, and seem to be giving good results. Increased runoff abetting floods is caused by expanding urban areas, and the effects are worsened by towns having outgrown the original sewer/storm drain system which cannot cope with the increased input to them.

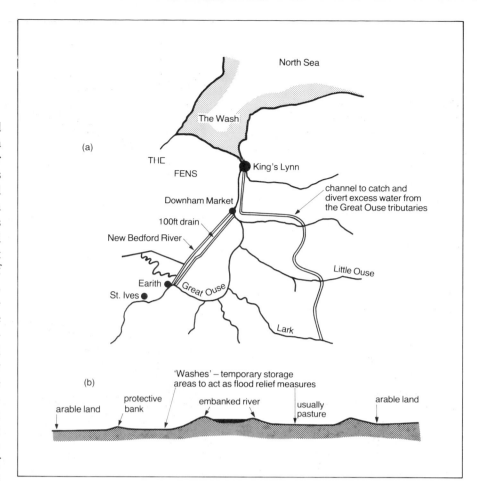

Fig 8.13 (a) the flood relief system of the Great Ouse (b) Fenland protective measures against floods or embankment failure

Fig 8.14 Man-made levées on the Mississippi

It is hard to see how this situation can be improved. New drain systems are part of the answer, but to attempt to delay runoff from urban areas is not compatible with reducing urban flood risk.

More direct types of flood control with a greater impact on the landscape include large-scale engineering projects. These include dams, river diversions, enlarging channel capacity and building protective embankments.

Dams are, arguably, the least effective method of flood control, and certainly they have a large number of problems. Dams need to be sited as closely as possible to the features requiring protection because their effect diminishes downstream, but dam construction in lower valleys consumes good land, and building smaller dams in the higher courses of rivers further removes flood protection from those areas which are most in need of protection. In many rivers the best dam sites have already been used so that future protection cannot be guaranteed at the same rate, though of course it may not be as much needed. Dams are also very costly, which means that they cannot be used only for flood control. Multi-purpose dam development, admirable in intent, faces conflict between the different groups who are hoping to use it. The summer water-skier and angler demand high water levels when these need to be lowered to meet the needs of irrigation; dams need to be empty to give maximum flood protection and full for maximum power generating and visual effect; maintaining a constant channel depth downstream for navigation is another matter again. The construction of dams can upset the equilibrium and ecology of downsteam channels both by impounding silt and reductions in flow volume. Lastly, there is always the prospect of dam failure.

River diversions fall into two broad categories, those which create a permanent new course for the river, and those which make provision for temporary diversion in times of peak flood. Great Britain has examples of both in East Anglia (see fig 8.13). The River Ouse has been diverted by the Bedford River and a hundred-foot drain to the west of Ely, while to the east of the town a diversion channel from near Mildenhall via Downham Market to King's Lynn intercepts high discharges from the Ouse's eastern tributaries. In addition to these measures the rivers and diversions of the Fens have 'washes', which are areas of protected flat ground adjacent to them which can be flooded as relief storage areas when floods threaten (see fig 8.13). Between times the washes contain pasture whose quality is enhanced by addition of silt from the flood waters.

On a very much larger scale are the Soviet Union's schemes to reduce flooding in the Arctic North by diverting northward-flowing rivers southwards into more arid parts of the country thereby solving two problems.

Enlarging channel capacity to accommodate flood waters is a popular

Fig 8.15   The Thames barrage at Greenwich

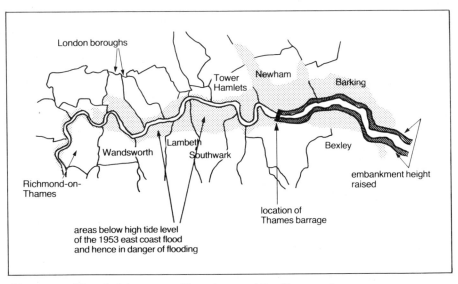

Fig 8.16   Flood-risk areas of London and the Thames barrage

flood prevention method but it is not without problems. The enlarged channel will disturb the river's equilibrium and will tend to silt up requiring constant dredging to prevent the channel reverting to its equilibrium form. Local water tables may also be affected, reducing the amount of water available to agriculture. In addition, enlarged channels are often straightened and their bed smoothed to ensure a more rapid flow, but unless this forms part of an integrated basin scheme the problem of the flood is merely passed

downstream – a clear case of having to approach the entire drainage basin as one unit.

The most common method of flood protection for rich and poor countries alike is building river embankments but, despite the fact that the Mississippi alone has nearly 4000km of them, they are not without their problems (see fig 8.14). Embankments resemble the natural levees created by rivers but in urban areas the cost of land which they would consume necessitates that they are built of steel or concrete. There is a tendency for aggradation of the embanked channel to take place requiring ever higher embankments which increase the consequences of embankment failure to adjacent land. Such failure is unlikely in modern embankments whose material is chosen for its cohesiveness and lack of permeability; sometimes such material may not be available locally and have to be imported, raising the cost of the project. Man-made levees are not 'natural' and by preventing floods may reduce the amount of water stored in a drainage basin, and even cause *dessication*, which is shrinkage of nearby land, particularly if it has a high peat content. Tributaries present great problems if the main stream is embanked, because the embankments act as dams causing the tributaries to flood. Solutions are costly, either embanking the tributaries too, or building sluices and pumps into the main stream. Lastly, bearing in mind the concept of the Possible Maximum Flood, embankments, once breached, make the disaster even greater and may prevent receding flood water draining back into the main stream.

Protection against coastal floods is done in two ways, by constructing sea walls on exposed coastlines and building barrages on indented coasts. In Britain sea walls exist in a number of localities, notably The Wash, but estuarine barrages are limited to the one nearing completion on the Thames at Greenwich (figs 8.15 and 8.16). Excellent examples of both types of control are found in the Netherlands whose famous saying 'God made the land but man made Holland' is true for the 30 per cent of the country reclaimed but below sea level. The Dutch have reclaimed river polders, lake polders, and polders from the sea (fig 8.17). They are currently engaged on an extremely ambitious three-stage programme to reclaim more land from the sea. Stage one, the reclamation of the Zuider Zee (now Lake Yjssel) is nearly completed. Stage two, reclamation and flood protection of the great Rhine delta is well under way. Stage three, the idea of joining the Fresian Islands to each other and the mainland, is still being planned, but in view of the fact that work on the last of Zuider Zee polders has not been started

because of huge environmental opposition, it could well be that stage three will have to await equally strong pressure on, and demand for land before being started in the future.

There are also plans to build barrages in the United Kingdom. Locations proposed include The Wash, Morecombe Bay, Solway Firth and the estuaries of the Severn, Dee and Humber. All schemes would be multi-purpose, the benefits of flood control, recreation, power generation, improved communications and so on, being weighed against problems of navigation, ecology including fish migration, as well as effects on the coastal geomorphology of adjacent areas (fig 8.18).

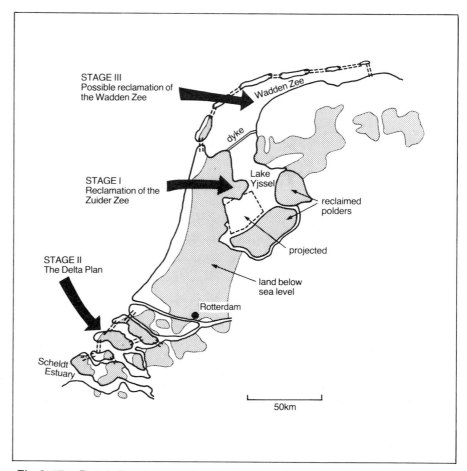

Fig 8.17   Dutch flood protection and land reclamation schemes

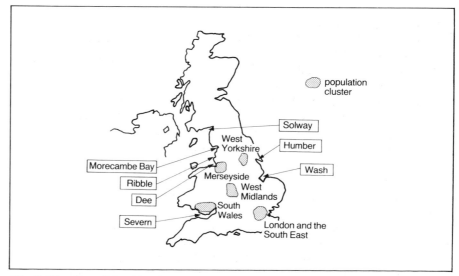

Fig 8.18   Possible sites for estuarine barrage locations in the United Kingdom, and major population clusters

EXERCISES

8.1 Go and look at a low-lying area near your home or school. What would you want to know in order to assess whether people living in the area should pay extra insurance premiums because of flood risk, and how high the premiums would be?

8.2 Study the Chew Valley lake system (fig 8.3) and try to produce a similar diagram for either
(i) another lowland reservoir in Britain or
(ii) an upland reservoir in Britain. Be as specific as you can about source of supply, storage site and nature of demand.

8.3 Consider figure 8.19 which shows Severnside, a site proposed for an estuarine barrage. Consult your atlas and library books for extra information.
(i) Why do you think Severnside has been considered as a site for a barrage?
(ii) Write a balanced view of the advantages and disadvantages of the location. You should consider extra water supplies (for whom?), improved road links (where to, and why are they needed?), land reclamation (for agriculture or industry?), a better environment for fish and fowl (would it be?), changes in tidal pattern, facilities for shipping and tourism.
(iii) Is Severnside a better estuarine barrage site than any other in Britain?

8.4 Write an essay on one of the following topics:
(i) When and why do floods occur?
(ii) Why does man live in areas liable to flood?
(iii) How does man plan for and respond to floods?

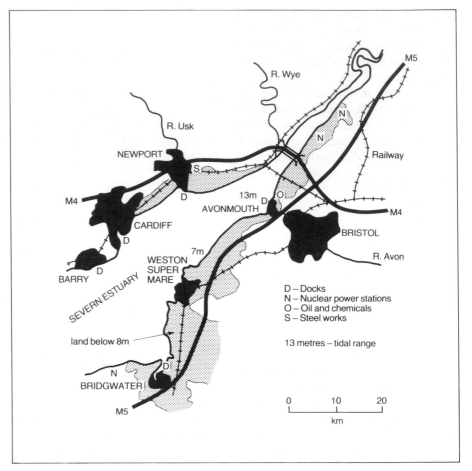

Fig 8.19   Severnside

9 Applied hydrology: world case studies

All societies make demands on the hydrological cycle. The ease with which they do so reflects their affluence and technological abilities. In part, deficiencies in these respects can be overcome by traditional expertise and cultural adaptation but, as the following case studies indicate, water management is often linked to stage of development and access to money.

The Tennessee Valley Authority

The TVA is the most famous example of water control in the world because it pioneered, in 1933, a large-scale multi-purpose development which has served since then as the model for similar schemes the world over.

The Tennessee River and its tributaries drain an area larger than the Benelux countries of Europe; the main river is approximately 1287km long, rising in the Blue Mountains of Virginia before flowing into the Ohio River shortly before the latter enters the mighty Mississippi; parts of seven states are within its catchment area which totals some 103 200km² (see fig 9.1).

The physical background of the scheme illustrated many hydrological issues. Prior to its control the Tennessee River flooded regularly, damaging farmland in its own valley and causing floods in the lower Ohio and Mississippi downstream beyond Memphis, over 1609km from its source. The river's regime was very irregular because deforestation of its catchment area, particularly of the steep slopes near its source, in combination with thoughtless farming, reduced infiltration, and caused rainfall to run off rapidly into the river systems as quickflow. This surface runoff caused severe soil erosion, soil which, between floods choked the river channels, making them useless to navigation. There was no water management in operation.

In part these conditions stemmed from, and caused, the social conditions which existed. The valley contained 2 million people, 85 per cent whites, many of whom lived relatively poor lives in the poverty and despair which characterised much of rural America during the 1920s and 1930s. Incomes were extremely low with no prospect of

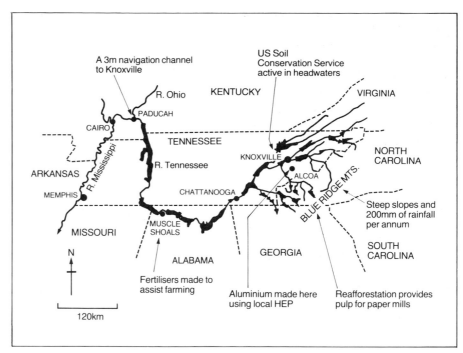

Fig 9.1   The Tennessee Valley Authority

non-agricultural employment because industry shunned the area.

Transformation came with Franklin Roosevelt's Democratic Party 'New Deal' for Americans, and the large amounts of Federal capital poured into the scheme represented a political and socio-economic investment. Firstly the river was controlled by building and regulating over thirty dams, small ones in the hilly source areas for runoff control, and larger multi-purpose ones further downstream. Education and welfare services raised living standards not least by improving health by virtually eliminating malaria, and agricultural advisory services were set up. Contour ploughing, damming gullies with brushwood, crop rotation and afforestation not only reduced soil erosion by improving infiltration, but also elevated the quality of the landscape sufficiently to attract tourists. The hydro-electric power generated by the dams, linked to the reliable 2.7m navigable channel from the Ohio to Knoxville attracted a variety of industry which has stimulated industry outside the immediate area, into much of the southern United States. The scheme, despite Republican opposition, has proved an immense success, not least in that it persuaded seven states to co-operate to bring a sense of unity and purpose to a depressed area of the United States.

Fig 9.2   Rain-making in California!

Water in California

Southern California's mediterranean climate has proved irresistible to specialised farmers, in latter years to a variety of industry, and to people from elsewhere in the United States migrating either for retirement, or to the better climate the area offers. All this has necessitated considerable efforts to meet the ever-increasing demands for water in the state.

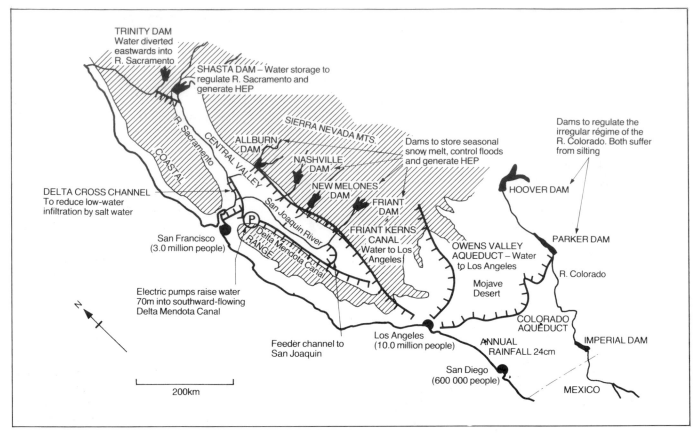

Fig 9.3   The Californian Central Valley Project

The original Indians dry-farmed and collected some snow-melt water for their crops, and it was left to the Spanish settlers to improve water supplies by introducing small earth-work dams and clay-pipe irrigation which their forefathers had used in Spain. A hundred years later in the mid-nineteenth century California's artesian basin was tapped for the first time by deep wells and steam pumps; by 1900 concrete irrigation pipes, electric pumps, dams and aqueducts had been introduced. The first half of the twentieth century saw prolific diversification of agriculture based on improved transport, refrigeration and the steadily increasing provision of water. After World War II the population of the state increased rapidly; Los Angeles grew by 200 000 a year between 1950–60; San Diego by 70 per cent in the same period as industry invaded the state, and because California represented a marvellous outdoor playground for its people.

All this development has had repercussions on water provision in the state. The agricutural demand in the citrus-growing Central Valley has led to a steadily falling water table as abstraction exceeds recharge. Well capture has been common and the future of the aquifer threatened as sea-water moves into strata which hitherto

it had been prevented from entering by the presence of fresh water. To try to alleviate the situation there has been resort to rain-making (see fig 9.2), but the most effective schemes have used long-distance water transfer.

Southern California is hot and dry; it contains major industrial and domestic demands for water (the latter often for watering lawns) while Northern California is cool, wet, with rainfall exceeding local demands. The obvious solution is to transfer water from the wet north to the dry south. The Central Valley project (see fig 9.3) does this using 611km of natural and man-made channels. Even this scheme is not enough to meet the demands of Southern California and the Owens Valley aqueduct brings water to Los Angeles from the Sierra Nevada Mountains. A third element of the plan is the Colorado. In a major scheme this river has been dammed by the Parker and Hoover dams, from which water is led across the Mojave Desert by the Colorado Aqueduct to Los Angeles. What little Colorado water escapes these dams is retained by the Imperial dam for irrigating the Imperial and Coachella valleys. Very little Colorado water reaches the sea via its natural channel, and the river provides a good example of man's modification of the hydrological cycle.

Towards a Nile valley authority

The Nile is the longest river in the world, and the difficulties which explorers like Burton and Speke had in finding its source indicate the likely problems in developing it as a functioning unit in the manner of the TVA. Countries within the drainage basin are poor, and would benefit from power generation, flood control, and the possibilities of extending irrigation that might come about. Some schemes have already been established, particularly in the more affluent lower reaches where international financial and technical assistance have been more readily available, but in general lack of money, political mistrust and a general air of procrastination have left many schemes for a Nile valley authority in abeyance.

Figure 9.4 shows the schemes which already exist, and the projected schemes (P) for the fuller control of the river. Most of the completed schemes are in the lower course of the river; control would be better effected by harnessing the rivers near their source. The British-built Owen Falls Dam on Lake Victoria generates large amounts of hydro-electric power for Uganda's textile industry as well as stabilising the water levels in the lake. Similar schemes

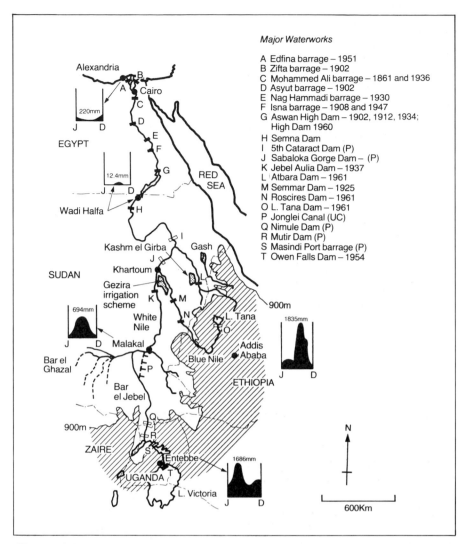

Fig 9.4   The Nile basin: major rivers, relief, selected rainfall totals and major water projects

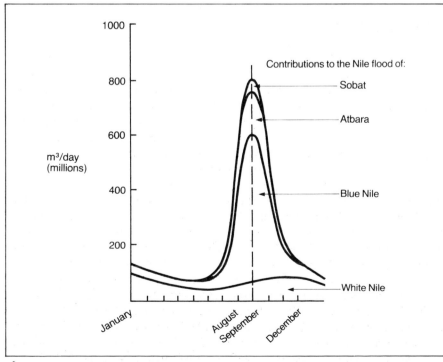

Fig 9.5   The Nile hydrograph

on Lake Kioga and Lake Albert might have similar effects and also assist navigation as far north as the Sudd. The Sudd is the key to much of the Nile development; its papyrus swamp absorbs vast amounts of salt-laden water, swelling and decreasing from wet to dry season, ensuring that clear water leaves Malakal as the White Nile. The Jonglei Canal, a scheme to by-pass the swamp, would increase the northerly movement of water by the amount lost in evaporation from the Sudd, and assist in navigation through the swamp, though its ecological changes would affect both man and beast, perhaps to a large degree.

The second major development to the Nile basin would be the Lake Tana Dam. The Blue Nile accounts for most of the Nile flood (see fig 9.5). It already has dams at Roseires and Sennar, and a dam on Lake Tana would increase control of the Blue Nile and bring power to impoverished Ethiopia. Other possible schemes would exert similar control on the Sobat and Atbara rivers, and improve navigation – and hence trade – along the length of the Nile by building dams at the famous cataracts to produce energy, and locks to by-pass the rapids.

Some of the benefits of water control are already evident along the Nile. The irrigation at Gezira, Gash and Kashm el Girba has produced some of the wealthiest peasantry in Africa, while the Aswan Dam, despite its problems, has increased Egypt's irrigated area by one third, generated power to pump water further into the densely-settled banks of the Nile, and become a symbol of Egypt's development.

Irrigation

Irrigation is a method of maintaining soil moisture levels sufficient to meet the demands of plants for water. Plants need water for four main reasons:
(1) to absorb minerals in solution;
(2) to enable photosynthesis to take place;
(3) to maintain plant temperature at a level at which the plant can function;
(4) to maintain turgidity;
and three states of water are of significance to plants:
(i) soil at field capacity, i.e. saturated;

(ii) soil whose water content has fallen to permanent plant wilting point;
(iii) soil which is totally dry.

Irrigation aims to keep moisture conditions between (i) and (ii) and gives irrigated crops four main benefits:
(1) a degree of immunity from climatic unreliability;
(2) increased yields;
(3) improved quality crops;
(4) earlier marketing potential.

In this sense irrigation is an agricultural asset, a technique to intensify land use and not a climatological necessity though, with increasing development of the world's arid lands, it is likely to become even more important. Since 1960 the world's irrigated area has increased by 10 per cent per annum.

Irrigation is an ancient practice, and there is evidence of its use in the earliest civilisations, often showing a surprising degree of social investment and organisation. Although basin irrigation, shadufs, saqias and qanats are still present in many parts of the world, modern emphasis has shifted towards large dams and concrete conduits; tube wells and sprinkler systems. Some of the problems have changed too.

Irrigation in the Indus valley

The British initiated many irrigation schemes in the Indian sub-continent. One with far-reaching effects began on the Indus plains at the turn of the century. The plain is almost perfectly flat, and the Indus, although slightly seasonal because of the monsoon, provides large amounts of water. Since the first barrage at Sukkur in 1932 (see fig 9.6), four others have been built to raise the water level so that it can be led further across the plains by permanent and seasonal canals from which it can be fed to the small ditches which supply the earth-banked fields and allowed to soak in. The system seems ideal and over 30 million people, 75 per cent of them farmers, depend on it to prevent the desert conditions found in nearby Bihar.

Unfortunately the canals leak, and there is too small a gradient to permit adequate drainage. Consequently the water table has steadily risen until it is too close to, or at, the surface over much of the area. Crops are killed by either water-logging or by saline crusts

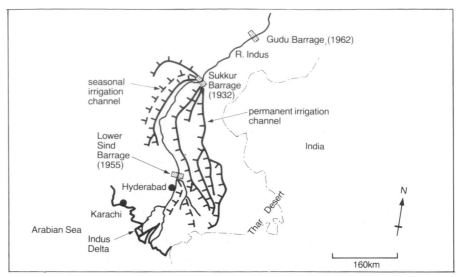

Fig 9.6 *Major irrigation channels of Sind Province, lower Indus valley, Pakistan*

Fig 9.7 *Soils ruined by salt crusts*

which follow the rapid evaporation of water at the surface (see fig 9.7). The problems of this scheme are compounded by the political confrontation which has produced bellicose warnings of conflict between India and Pakistan throughout the life of the scheme. Only in 1960 was the Indus Water Treaty signed guaranteeing supplies. Remedies to the former problem revolve around lowering the water table by drilling deep tube wells and pumping water away to be used for irrigation elsewhere, a form of vertical drainage which, it is hoped, will leach the surface salts away. The latter problem which cannot be solved by mechanical or technological means, provides another significant example of what is one of the major issues facing hydrologists today – that of overcoming the political difficulties of drainage basin management in an increasingly divided world.

EXERCISE

9.1 Look up in your library examples of water management schemes. Make notes on the schemes, paying strict attention to the economic, social as well as the hydrological background.

Some examples could be, for the developed world: the Snowy Mountain scheme of Australia, the Vaal-Harts scheme of South Africa, the Red River area of the United States. For the less developed world: the Gezira Project on the River Nile, the Sansanding scheme on the River Niger, the Kariba and Cabora Bassa schemes on the River Zambezi.
(i) What techniques did the ancient Egyptians use to irrigate their lands?
(ii) Why was the Aswan Dam built, and what problems have been encountered because of it?

Booklist

1 D. Briggs, *Soils*, Butterworth
2 British Geomorphological Group Technical Bulletin No 4, 'Drainage Basin Morphometry'
3 R.J. Chorley, *Water, Earth and Man*, Methuen
4 K. Hilton, *Process and Pattern in Physical Geography*, UTP
5 J.W. House (Ed.), *The U.K. Space*, Weidenfeld and Nicolson
6 S. Jones and R.A. Beddis, *Water in Britain*, Hodder and Stoughton
7 L. Leopold, M. Wolman and J. Miller, *Fluvial Processes on Geomorphology*, W.H. Freeman and Co.
8 P. McCullagh, *Modern Concepts in Geomorphology*, Oxford University Press
9 Marie Morisawa, *Streams: Their Dynamics and Morphology*, McGraw-Hill
10 M.D. Newson, *Flooding and Flood Hazard in the U.K.*, Oxford University Press
11 J. Oliver, *Perspectives on Applied Physical Geography*, Duxbury
12 D.V. Parker and E.C. Penning-Rowsell, *Water Planning in Britain*, George Allen and Unwin
13 H.C. Pereira, *Land Use and Water Resources*, Cambridge University Press
14 A.F. Pitty (Ed.), *Geomorphological Approaches in Fluvial Processes*, Geo. Books
15 M. Simons, *Deserts*, Oxford University Press
16 D.I. Smith and P. Stopp, *The River Basin*, Cambridge University Press
17 K. Smith, *Water in Britain*, Macmillan
18 K. Smith and G.A. Tobin, *Human Adjustment to the Flood Hazard*, Longmans
19 *Discharge of Selected Rivers of the World*, UNESCO
20 R.C. Ward, *Principles of Hydrology*, McGraw-Hill
21 R.C. Ward, *Floods*, Macmillan
22 D.R. Weyman, *Run-off Processes and Stream-flow Modelling*, Oxford University Press

Acknowledgements

The publishers are grateful to the following for providing photographs:

Aerofilms: 5.3, 5.9, 6.5, 8.14
Associated Press Ltd: 8.1, 8.8
N. Barrington: 8.2
The Forestry Commission: 2.5
Gloucestershire Newspapers Ltd: 8.5
The Greater London Council: 8.15
The Ministry of Information, Zimbabwe/Harrison Church: 1.8
National Oceanic and Atmospheric Administration/National Environmental Satellite Service/Science Photo Library: 8.7
M.D. Newson: 2.4, 2.8, 4.6, 4.16
Oxford Scientific Films (John Paling): 9.7
Popperfoto: 2.1, 2.6
Rida Photo Library: 5.10
Ronald Sheridan Photo Library: 3.3
Soil Survey of England and Wales, Rothamsted Experimental Station: 2.12
South West Water Authority: 1.2
Wessex Water Authority: 1.1, 1.3
York and County Press (Westminster Press Ltd): 8.11

Index